THE PROFESSIONAL ENGINEERING CAREER DEVELOPMENT SERIES

CONSULTING EDITOR

Mr. Dean E. Griffith, Director, Continuing Engineering Studies, The University of Texas at Austin

CONSULTANTS

Dr. John J. McKetta, Jr., E. P. Schoch Professor of Chemical Engineering, The University of Texas

Dr. Maurits Dekker

EDITORIAL ADVISORY COMMITTEE

Dr. Walter O. Carlson, Acting Dean of Engineering College, Georgia Institute of Technology

Mr. R. E. Carroll, Director, Continuing Engineering Education, The University of Michigan

Mr. William W. Ellis, Director, Post College Professional Education, Carnegie-Mellon University

Dr. Gerald L. Esterson, Director, Division of Continuing Professional Education, Washington University

Dr. L. Dale Harris, Associate Dean, College of Engineering, The University of Utah

Dr. James E. Holte, Director, Continuing Education in Engineering and Science, University of Minnesota

Dr. Russell R. O'Neill, Associate Dean and Professor, School of Engineering and Applied Science, University of California at Los Angeles

D1127933

ENGINEERING PROFESSION ADVISORY GROUP

Professional Engineering Career Development Series

FUNDAMENTAL PRINCIPLES OF POLYMERIC MATERIALS FOR PRACTICING ENGINEERS

Stephen L. Rosen

Department of Chemical Engineering
Carnegie-Mellon University
Pittsburgh, Pennsylvania

BARNES & NOBLE, INC. NEW YORK

Publishers • Booksellers • Since 1873

L. C. Catalogue Card Number: 76-146260

ISBN 389 00511 8

Distributed

In Canada
by the Ryerson Press, Toronto

In Australia and New Zealand
by Hicks, Smith & Sons Pty. Ltd.,
Sydney and Wellington

In the United Kingdom, Europe,
and South Africa
by Chapman & Hall Ltd., London

In Japan
by Maruzen Co. Ltd., Tokyo

Printed in the United States of America

Preface

The commercial growth of polymers after World War II was so rapid that technology far outdistanced the fundamental knowledge on which it is ultimately based. Much of the blame for this situation must lie with the academic world which, with a few notable exceptions, abdicated its responsibility to provide graduates trained in these fundamentals. Academic chemists refused to study "gunk," materials scientists wouldn't acknowledge the existence of materials with less than 99.9 percent crystallinity, civil and mechanical engineers didn't develop design procedures to handle materials with time-dependent properties, and chemical engineering educators didn't recognize that nearly half their graduates would ultimately wind up working with polymers in one way or another.

Fortunately, the situation has changed for the better in the last decade, and the fundamental knowledge is catching up with technology. The continued profitable growth of polymer-based industries rests on three factors: (1) in the case of older, high-volume, low-profit polymers, the edge will go to the product which is slightly better suited, through production, processing and design, to its use and to the organization which can provide better technical service for its customers; (2) the development of new polymers with properties and prices superior to existing materials; and (3) the development of new applications for existing polymers. It is clear that engineers with a knowledge of the fundamentals are necessary in each area.

The object of this book is to provide an appreciation of those fundamental principles of polymer science and engineering which are currently of practical relevance. It is hoped that the reader will obtain (1) a broad, unified introduction to the subject matter which will be of immediate practical value, and

(2) a foundation for more advanced study, either in the current literature or in specialized courses.

Obviously, the choice of material to be covered involves subjective judgement on the part of the author and, together with space limitations, necessitates omission or shallow treatment of many interesting developments. The References and Selected Readings are specifically chosen to aid the reader who may wish to pursue a subject in greater detail.

In all cases, primary importance is placed on a qualitative understanding of the underlying principles before tackling any mathematical developments. Particular stress is placed on how a polymer's molecular structure determines its processing and product properties and how the desired structure is obtained in the synthesis step. Section 2, "Polymer Synthesis," will be of greatest interest to chemists and chemical engineers; it may be omitted by others, but perhaps it will at least prove of cultural interest to them, also,

Acknowledgments

The author would like to sincerely thank the following: his students, for proving time and again that the best way to learn is to teach; his teachers, Professors B. Maxwell, L. Rahm, H. Pohl and A. V. Tobolsky and, in particular, Professor Ferdinand Rodriguez, for making the "macromolecular gospel" so interesting (any similarities between this book and Professor Rodriguez' recent, excellent text are a direct result of many fascinating hours in his classes and laboratory); his many industrial friends and colleagues, for keeping him aware of the important "real world" problems; his wife, for the many hours she spent alone in front of the "boob tube" as this was being written; Mrs. Dolores Dlugokecki, for somehow maintaining her cheerful sanity while typing the manuscript.

Contents

SECTION 3
MECHANICAL PROPERTIES OF POLYMERS

SECTION 4
POLYMER TECHNOLOGY

FUNDAMENTAL PRINCIPLES OF POLYMERIC MATERIALS FOR PRACTICING ENGINEERS

1
Introduction

Since the Second World War, polymeric materials have been the fastest-growing segments of the United States chemical industry. It has been estimated that more than 25 percent of the chemical-research dollar is spent on polymers, with a correspondingly large proportion of technical personnel working in the area. As production climbs, so do applications. In the 1970-model year, plastics surpassed the 100-pound-per-car level, and this does not include paints and the rubber and fibers in tires. At nearly ten million cars per year, this is a tremendous market, and, as polymers continue to replace traditional materials, it promises to become even greater. Similarly, the applications of polymers in the building-construction industry (piping, resilient flooring, thermal and electrical insulation, paints, decorative laminates, etc.) are already impressive and will become even more so in the near future. Many other examples could be cited, but, to make a long story short, by around 1980 the use of polymers should outstrip that of metals on a volume basis.

There are five major areas of application for polymers: (1) plastics, (2) rubbers or elastomers, (3) fibers, (4) surface finishes and protective coatings, and (5) adhesives. Despite the fact that the five applications are all based on polymers (in many cases, the same polymer is used in two or more), the industries grew up pretty much separately. It was only after Dr. Herman Staudinger proposed the "macromolecular hypothesis" in the 1920s, explaining the common molecular makeup of these materials (for which he won the 1953 Nobel Prize in chemistry in belated recognition of the importance of his work), that polymer science began to evolve from the independent technologies. Thus, a sound fundamental basis was established for continued technological advances.

1

Economic considerations alone would be sufficient to justify the impressive scientific and technological efforts expended on polymers in the past several decades. In addition, however, this class of materials possesses many interesting and useful properties which are completely different from those of the more traditional engineering materials and which cannot be explained or handled in design situations by the traditional approaches. A description of three simple experiments should make this obvious.

"Silly putty," a silicone polymer, bounces like rubber when rolled into a ball and dropped. On the other hand, if the ball is placed on a table, it will gradually spread to a puddle. The material behaves as an elastic solid under certain conditions and as a very viscous liquid under others.

If a weight is suspended from a rubber band and the band is then heated (taking care not to burn it), the rubber band will *contract* appreciably. All materials other than polymers will undergo the expected thermal *expansion* upon heating (assuming no phase transformation has occurred over the temperature range).

When a rotating rod is immersed in a molten polymer or a fairly concentrated polymer solution, the liquid will actually climb up the rod. This phenomenon, the Weissenberg effect, is contrary to what is observed with nonpolymer liquids, which develop a parabolic surface profile with the lowest point at the rod as the material is flung outward by centrifugal force.

Although such behavior is unusual in terms of the more familiar materials, it is a perfectly logical consequence of the *molecular structure* of polymers. This molecular structure is the key to an understanding of the science and technology of polymers, and it will underlie the chapters to follow.

Figure 1.1 illustrates the questions to be considered:
1. How is the desired molecular structure obtained?
2. How do the polymer's processing properties depend on its molecular structure?
3. How do its material properties (mechanical, chemical, optical, etc.) depend on molecular structure?
4. How do material properties depend on a polymer's processing history?
5. How do its applications depend on its material properties?

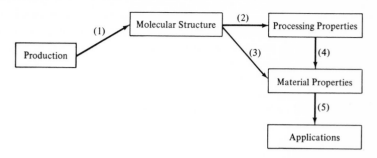

Figure 1.1. The key role of molecular structure in polymer science and technology.

The word 'polymer' comes from the Greek words 'many membered.' Strictly speaking, it could be applied to any large molecule which is formed from a relatively large number of smaller units (or mers)—e.g., a sodium chloride crystal; however, it is most commonly (and exclusively, here) restricted to materials in which the mers are held together by covalent bonding—i.e., shared electrons. For our purposes, only a few bond valences need to be remembered:

$$-\overset{|}{\underset{|}{C}}- \quad \overset{|}{\underset{/\,\backslash}{N}} \quad -O- \quad Cl- \quad F- \quad H- \quad -\overset{|}{\underset{|}{Si}}-.$$

It is always a good idea to count the bonds in any structure written to make sure they conform to the above. A brief, concise review of organic chemistry from the polymer standpoint is available (1).

The most important constituents of living organisms, cellulose and proteins, are naturally occurring polymers, but we will confine our attention largely to synthetic polymers or to important modifications of natural polymers.

REFERENCES

1. Richardson, P. N. and R. C. Kierstead. Organic chemistry for plastics engineers. *Soc. Plast. Eng. J.*, **25,** no. 9, p. 54, 1969.

Section 1

Polymer Fundamentals

2
Types of Polymers

The large number of natural and synthetic polymers have been classified in several ways. These will be outlined below, and, in the process, many terms important in polymer science and technology will be introduced.

2.1 REACTION TO TEMPERATURE

The earliest distinction between types of polymers was made long before any concrete knowledge of their molecular structure. It was a purely phenomenological distinction, based on their reaction to heating and cooling.

Thermoplastics It was noted that certain polymers would soften upon heating and could then be made to flow when a stress was applied. When cooled again, they would reversibly regain their solid or rubbery nature. These polymers are known as *thermoplastics*.

Thermosets Other polymers, although they might be heated to the point where they would soften and could be made to flow under stress *once*, would not do so reversibly; i.e., heating caused them to undergo a curing reaction. Sometimes these materials emerge from the synthesis reaction in a cured state. Further heating of these *thermosetting* polymers ultimately leads only to degradation (as often attested to by the smell of a short-circuited electrical appliance) and not softening and flow.

Continued heating of thermoplastics will also lead ultimately to degradation, but they will generally soften at temperatures below their degradation point.

4

A classic example of these two categories is natural rubber. Introduced to Europe by Columbus, natural rubber did not achieve any commercial significance for centuries because it was a thermoplastic—i.e., articles made of it would become soft and sticky on hot days. It was only when Goodyear discovered the curing reaction which converted the polymer to a thermoset—which he called vulcanization in honor of the Roman god of fire—allowing it to maintain its useful rubbery properties to much higher temperatures that rubber began to achieve commercial importance.

2.2 CHEMISTRY OF SYNTHESIS

Pioneering workers in the field of polymer chemistry soon observed that they could produce polymers by two familiar types of organic reactions.

Condensation Polymers formed from a typical organic condensation reaction, in which a small molecule (most often water) is split out, are known, logically enough, as *condensation polymers*. The common esterification reaction of an organic acid and an organic base (alcohol) illustrates the simple "lasso chemistry" involved:

$$R-O{H} + {HO}-\overset{\displaystyle O}{\overset{\|}{C}}-R' \rightarrow R-O-\overset{\displaystyle O}{\overset{\|}{C}}-R' + H_2O.$$

alcohol + acid → ester

Of course, the ester formed in the preceding reaction is not a polymer; i.e., we have only hooked up two small molecules, and the reaction is finished far short of anything which might be considered many-membered.

At this point, it is useful to introduce the concept of *functionality*. *Functionality is the number of bonds a mer can form with other mers in a reaction*. It is obvious from the example above that each of the reactants is monofunctional and that reactions between monofunctional mers cannot lead to polymers. However, consider what happens if each reactant is *di*functional, allowing it to react at each end:

$$\text{HO—R—O\underline{H}} + \text{(HO)—C(=O)—R'—C(=O)—OH} \rightarrow$$

di alcohol   dicarboxyllic acid
or diol   or diacid

$$\text{HO—R—O—C(=O)—R'—C(=O)—OH} + H_2O.$$

The resulting product molecule is still difunctional; i.e., its left
end can react with another diacid molecule and its right end
with another molecule of diol, and, after each subsequent reaction, the growing molecule is still difunctional and capable of
undergoing further growth, leading to a true polymer molecule.

In general, the *polycondensation* of x moles of diol with x
moles of a diacid to give a *polyester* is

$$x\,\text{HO—R—OH} + x\,\text{HO—C(=O)—R'—C(=O)—OH} \rightarrow$$

$$\text{H}\!\left[\text{O—R—O—C(=O)—R'—C(=O)}\right]_x\!\text{OH} + (2x-1)\,H_2O.$$

The $\left(\!\text{O—C(=O)}\!\right)$ linkage is characteristic of a polyester. The

quantity x, the number of *repeating units* (enclosed in brackets)
in the polymer chain, is known as the *degree of polymerization*.

Another functional group which is capable of taking part in
a condensation is the ami*ne* (—NH$_2$) group, *one* hydrogen of
which reacts with a carboxyllic acid group in a manner similar
to the alcoholic hydrogen to form an ami*de* or nylon:

$$x\,\text{H}_2\text{N—R—N}\text{H}_2 + x\,\text{HO—C(=O)—R'—C(=O)—OH} \rightarrow$$

$$\text{H}\!\left[\text{N(H)—R—N(H)—C(=O)—R'—C(=O)}\right]_x\!\text{OH} + (2x-1)\,H_2O.$$

diamine + diacid $\rightarrow$ polyamide or nylon

The $\left(\!\!\begin{array}{c}H\ \ O\\ |\ \ \ \|\\ -N-C-\end{array}\!\!\right)$ linkage characterizes nylons.

The examples above serve to illustrate that reactants must be at least difunctional if a polymer is to be obtained. Molecules with higher degrees of functionality will also lead to polymers. For example, glycerine

$$H-\overset{\overset{\displaystyle H}{|}}{\underset{\underset{\displaystyle OH}{|}}{C}}-\overset{\overset{\displaystyle H}{|}}{\underset{\underset{\displaystyle OH}{|}}{C}}-\overset{\overset{\displaystyle H}{|}}{\underset{\underset{\displaystyle OH}{|}}{C}}-H$$ is trifunctional

in a polyesterification reaction.

In addition to using two monomers with the same functional group at each end, it is also possible to form condensation polymers from a single monomer containing the two complementary reactive groups in the same molecule:

$$HO-R-\overset{\overset{\displaystyle O}{\|}}{C}-OH \qquad \overset{\displaystyle H}{\underset{\displaystyle H}{>}}N-R-\overset{\overset{\displaystyle O}{\|}}{C}-OH.$$

hydroxy acid   amino acid

In principle, the hydroxy acid is capable of forming a polyester and the amino acid a polyamide (proteins are poly amino acids). The reactions do not always proceed in a straightforward fashion, however. If R is large enough (e.g., three carbon atoms or more), the difunctional monomers above may 'bite their own tails' and condense to form a cyclic structure:

$$\overset{\displaystyle H}{\underset{\displaystyle H}{>}}N-R-\overset{\overset{\displaystyle O}{\|}}{C}-OH \rightarrow \overset{\displaystyle R}{\underset{\displaystyle}{\Big\langle}}\begin{array}{c}\overset{\displaystyle O}{\|}\\ C\\ |\\ N\\ |\\ H\end{array} + H_2O.$$

amino acid   *lactam*

This cyclic compound can then undergo a *ring scission* polymerization, in which the polymer is formed without splitting out a small molecule, which had been eliminated previously in the cyclization step:

$$x \ R \underset{\substack{| \\ N \\ | \\ H}}{\overset{\substack{O \\ \| \\ C}}{\diagup}} \rightarrow \left[N - R - C \right]_x .$$

polyamide

Despite the lack of elimination of a small molecule in the actual polymerization step, the products can be thought of as being formed by a direct condensation from the monomer and are usually considered condensation polymers.

Before moving on, it is worthwhile to point out that the generalized organic groups, R, in the compounds above may vary widely. The choice of R can strongly influence the properties of the resulting condensation polymer.

Addition The second polymer-formation reaction is known as addition polymerization and its products are known as *addition polymers*. Addition polymerizations have two distinct characteristics:

1. no molecule is split out—hence, the repeating unit has the same formula as the monomer; and

2. the polymerization reaction involves the opening of a double bond. (Although aromatic rings are often symbolized

by ⬡, this is a poor representation of a resonance-stabilized structure which is completely inert to addition polymerization. They are more properly symbolized by ⬡ to avoid confusion with the ordinary double bond.)

Monomers of the general type $C = C$ undergo addition polymerization:

$$x \ C = C \rightarrow \left[C - C \right]_x ;$$

i.e., the double bond opens up, forming bonds to other monomers at each end, so *a double bond is difunctional*. The question of what happens at the ends of the polymer molecule will be deferred for a discussion of polymerization mechanisms.

An important subclass of the double-bond containing monomers are the vinyl monomers, $\overset{\overset{\displaystyle H}{|}}{\underset{\underset{\displaystyle H}{|}}{C}}=\overset{\overset{\displaystyle X}{|}}{\underset{\underset{\displaystyle H}{|}}{C}}$. Addition polymerization

is occasionally referred to as vinyl polymerization.

Example 1. Lactic acid can be dehydrated to acrylic acid according to the following reaction:

$$HO-\overset{\overset{\displaystyle H}{|}}{\underset{\underset{\displaystyle H}{|}}{C}}-\overset{\overset{\displaystyle H}{|}}{\underset{\underset{\displaystyle H}{|}}{C}}-\overset{\overset{\displaystyle O}{\|}}{C}-OH \;\rightarrow\; \overset{\overset{\displaystyle H}{|}}{\underset{\underset{\displaystyle H}{|}}{C}}=\overset{\overset{\displaystyle H}{|}}{C}-\overset{\overset{\displaystyle O}{\|}}{C}-OH \;+\; H_2O.$$

lactic acid   acrylic acid

Each acid forms a polymer. Write the structural formulae for the repeating unit of each polymer.

Solution. Lactic acid is an hydroxy acid and will undergo condensation polymerization, splitting out water from the

$-OH$ and $-\overset{\overset{\displaystyle O}{\|}}{C}-OH$ groups to give a polyester

$$\left[O-\overset{\overset{\displaystyle H}{|}}{\underset{\underset{\displaystyle H}{|}}{C}}-\overset{\overset{\displaystyle H}{|}}{\underset{\underset{\displaystyle H}{|}}{C}}-\overset{\overset{\displaystyle O}{\|}}{C}\right]_x.$$

Note that, in the above formula, the characteristic polyester

linkage $\left(O-\overset{\overset{\displaystyle O}{\|}}{C}\right)$ is split up. This illustrates that the location of the brackets denoting the repeating units is arbitrary.

Acrylic acid is a vinyl monomer and will undergo addition polymerization to give

$$\left[\overset{\overset{\displaystyle H}{|}}{\underset{\underset{\displaystyle H}{|}}{C}}-\overset{\overset{\displaystyle H}{|}}{\underset{\underset{\displaystyle C=O}{|}}{C}}\right]_x.$$
$$\underset{\underset{\displaystyle H}{|}}{\underset{\displaystyle O}{}}$$

Also capable of undergoing addition polymerization are the *dienes* (the carbon atoms in which are numbered from one end):

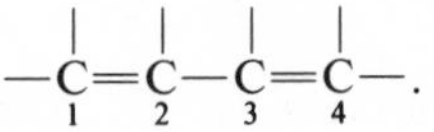

$$-\overset{|}{\underset{1}{C}}=\overset{|}{\underset{2}{C}}-\overset{|}{\underset{3}{C}}=\overset{|}{\underset{4}{C}}-.$$

Addition polymerization of dienes results in an *unsaturated* polymer—i.e., a chain which contains double bonds. Furthermore, there are several possibilities for the addition reaction. If the monomer is symmetrical with regard to substituent groups, it can undergo 1,2 addition and 1,4 addition. If unsymmetrical, there is the added possibility of 3,4 addition. (For symmetrical dienes, the 1,2 and 3,4 reactions are the same.) This is illustrated below for the addition polymerization of isoprene (3-methyl 1,3 butadiene).

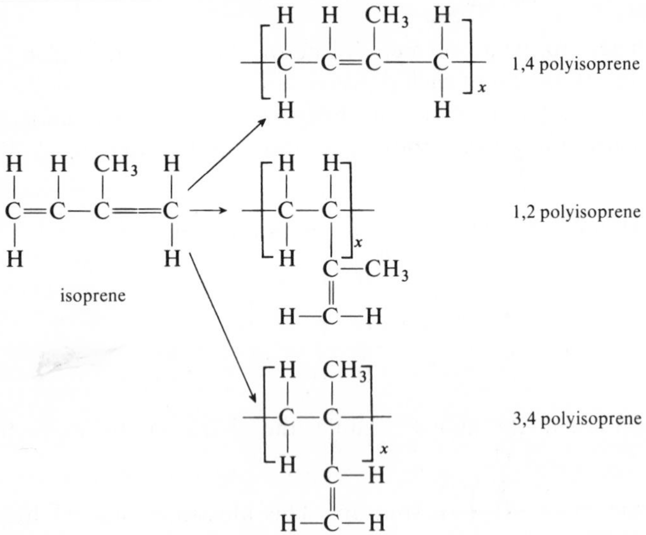

A wide variety of substituents is possible for the carbon atoms in the generalized monomers above, each conferring certain characteristics on the resulting polymer.

Oxidative Coupling Although presently used to produce only one commercial polymer, the oxidative coupling polymerization reaction is included here because of its potential. Oxidative coupling is applicable to certain monomer molecules containing two active hydrogen atoms:

$$x\,H—R—H + \frac{x}{2}\,O_2 \rightarrow -[R]_{\overline{x}} + x\,H_2O.$$

Commercially, 2,6 dimethylphenol is polymerized to poly-
phenylene oxide in the presence of a metal complex catalyst:

$$x\,HO-\underset{CH_3}{\overset{CH_3}{\bigcirc}}+\frac{x}{2}\,O_2 \rightarrow \left[O-\underset{CH_3}{\overset{CH_3}{\bigcirc}}\right]_x + x\,H_2O.$$

2.3 STRUCTURE

As an appreciation for the molecular structure of polymers
was gained, three major structural categories emerged.

Linear   If a polymer is built from strictly *difunctional*
monomers, the result is a *linear* polymer chain.   The term
linear is somewhat misleading, however, because the molecules
are never stretched out in a truly linear fashion.   In general,
an isolated linear molecule (e.g., in a dilute solution) assumes
a more or less random twisted and tangled configuration when
not subjected to an external stress.   Mathematical treatments of
such flexible chains, in fact, are based on the random walk
problem in three dimensions; i.e., if a drunk is turned loose
at a lamppost, with the direction of each of his steps being
completely random, where will he be after x steps (if you can
imagine him staggering in three dimensions)?   In practice,
polymer molecules are never isolated, of course, and the many
chains in a bulk sample are twisted and tangled together.   A
common analogy is a bowl of cooked spaghetti.   A far better
analogy, in view of the constant thermal motion of the chain
segments (for those with a strong stomach), is a bowl of
wriggling worms.

1. Random copolymers.   If, instead of starting out with a
pure monomer and polymerizing it by addition to form a
homopolymer, a mixture of two difunctional monomers A and
B is used, the resulting linear polymer will often have a random
arrangement of monomers A and B in the chain (the degree
of randomness depends on the relative amounts and reactivities
of A and B, as will be seen later).   The resulting polymer is
known as a *random copolymer* or just plain *copolymer*:

ABBBAABAAABBABAABBB.

Of course ter- and higher multipolymers are possible. The products of condensation polymerizations which require two different monomers to provide the necessary functional groups (e.g., a diacid and a diamine) are not copolymers. If, however, two different diamines were to be used, leading to two distinct repeating units, the product would be considered a copolymer.

2. Block copolymers. Under certain conditions, two (or more) monomers can be polymerized in such a manner that long blocks of each monomer are combined in a single chain, a *block copolymer*:

AAAAA---------AAAABBBB----------BBBAAA-------AAAAA.

Branched   If a few points of tri (or higher) functionality are introduced (either intentionally or through side reactions) at random points along linear chains, *branched* molecules result. Branching can have a tremendous influence on the properties of polymers through steric (geometric) effects.

1. Graft copolymers. Under specialized conditions, branches of monomer B may be grafted to a backbone of linear A. This structure is known as a *graft copolymer*:

$$
\begin{array}{ccc}
& B^B & B^B \\
B^B & B^B \\
B & B \\
AAAAA\text{-----}A\text{------}A \\
B \\
B \\
B
\end{array}
$$

Crosslinked or Network   As the length and frequency of the branches on polymer chains increases, the probability that the branches will finally reach from chain to chain, connecting them together, becomes greater. When all the chains are finally connected together in three dimensions by these *crosslinks*, the entire polymer mass becomes one tremendous molecule, a *crosslinked* or *network* polymer. A bowling ball, for example, has a molecular weight on the order of 10^{28}, all the polymer in it being connected to form one molecule by crosslinks.

Crosslinked or network polymers may be formed in two ways: (1) by starting with reaction masses containing sufficient amounts of tri- or higher functional monomer, or (2) by chemically creating crosslinks between previously formed linear or branched molecules (curing). The latter is precisely what vulcanization does to natural rubber, and this fact serves to introduce the connection between the phenomenological reaction-to-temperature classification and the more fundamental concept of molecular structure. This important connection will be clarified through a discussion of bonding in polymers.

Example 2. Show how a vinyl monomer (e.g., styrene,

) may be used to crosslink a linear, unsaturated

polyester—i.e., a linear polyester with double bonds in the chain.

Solution. The vinyl monomer undergoes an ordinary addition with the double bonds in the linear chains.

unsaturated
linear chains   styrene

network structure

Example 3. Show the structural formulae of the repeating unit for each of the following polymers, and classify them according to structure (linear, branched, network) and chemistry of formation (condensation, addition). A few oddballs—not following the general rules—have been thrown in.

- A. polystyrene
- B. low (0.92) density polyethylene
- C. high (0.97) density polyethylene
- D. polyethylene terephthalate (Dacron, Mylar)
- E. nylon 6/6
- F. nylon 6
- G. Glyptal (glycerol + phthalic anhydride)
- H. polydivinyl benzene
- I. melamine—formaldehyde (Melmac, Formica)
- J. polytetrafluorethylene (Teflon TFE)
- K. vulcanized natural rubber
- L. polypropylene
- M. acetal (polyformaldehyde) (Delrin, Celcon)
- N. polycarbonate (Lexan, Merlon)
- O. epoxy or phenoxy
- P. polydimethyl siloxane (silicone rubber)   (Hint: polymerized in presence of H_2O)
- Q. polyurethane

Starting monomers are shown below.

A. $H-C{=}CH_2$

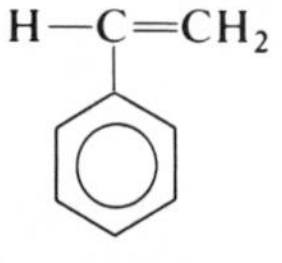

styrene

B. & C. $H_2C{=}CH_2$
ethylene

D. $H_3C-O-\underset{\underset{O}{\|}}{C}-\underset{}{\bigcirc}-\underset{\underset{O}{\|}}{C}-O-CH_3$

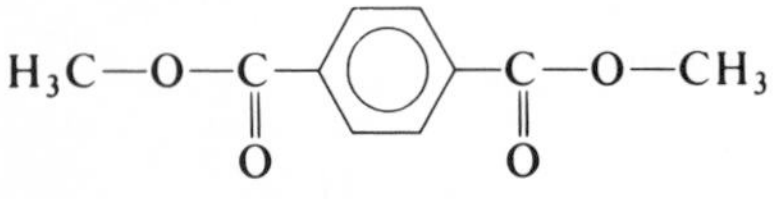

dimethyl terephthalate

$HO(CH_2)_2OH$
ethylene glycol

E. $H_2N(CH_2)_6NH_2$
hexamethylene diamine

$HO-\underset{\underset{O}{\|}}{C}-(CH_2)_4-\underset{\underset{O}{\|}}{C}-OH$
adipic acid

F.

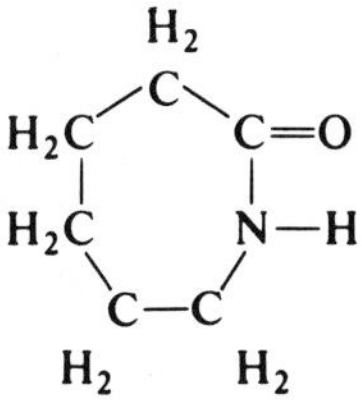

ε-caprolactam

G.

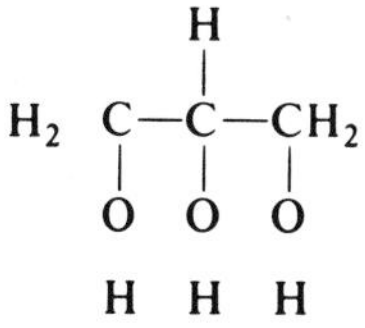

glycerol

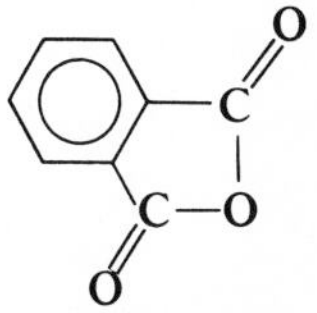

phthalic anhydride

H.

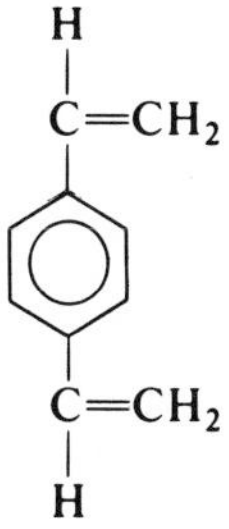

divinyl benzene

I.

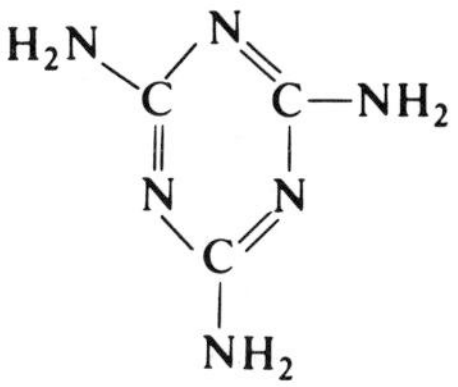

melamine

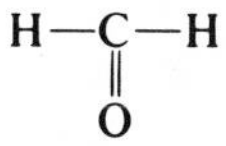

formaldehyde

J. $F_2C=CF_2$

tetrafluorethylene

K.

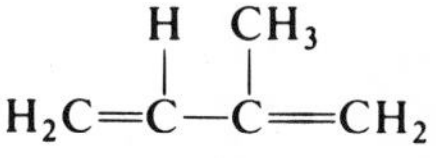

isoprene

L.

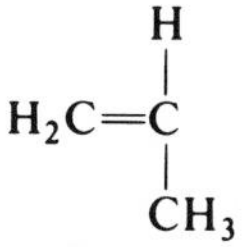

propylene

M.

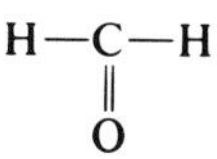

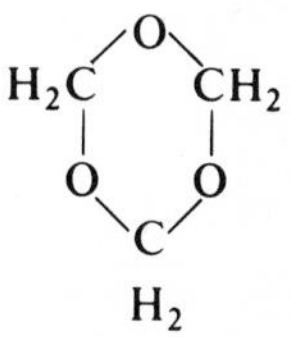

formaldehyde   trioxane

N. 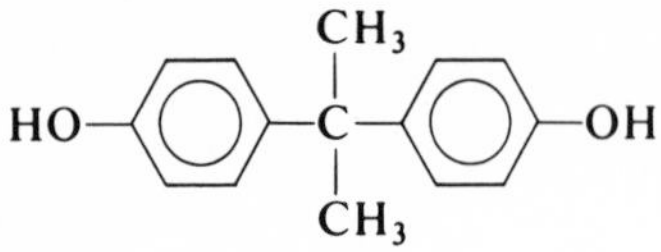 HO—⟨benzene⟩—C(CH₃)(CH₃)—⟨benzene⟩—OH

bisphenol-A

$$\text{Cl—C—Cl} \quad (\text{C}=\text{O})$$

phosgene

O. 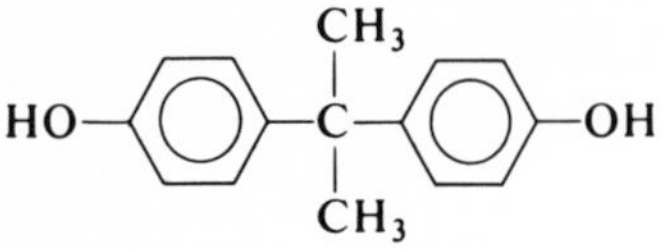 HO—⟨benzene⟩—C(CH₃)(CH₃)—⟨benzene⟩—OH

bisphenol-A

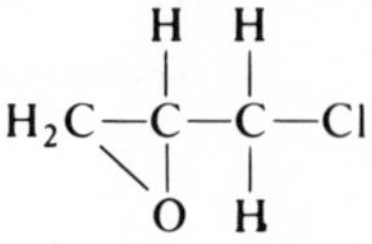

$$H_2C\text{—}\underset{O}{\overset{H}{C}}\text{—}\underset{H}{\overset{H}{C}}\text{—Cl}$$

epichlorohydrin

P. 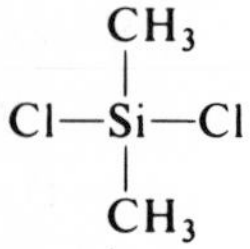   Cl—Si(CH₃)(CH₃)—Cl

dimethyl dichlorosilane

Q. **HOROH**

diol or glycol

$$O=C=N—R—N=C=O$$

diisocyanate

Solution

A. 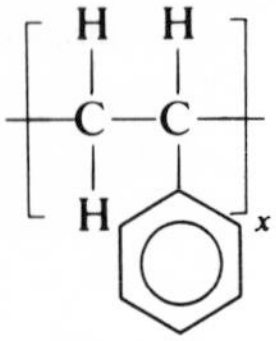 $\left[\begin{array}{c} \underset{H}{\overset{H}{C}}\text{—}\underset{\text{(phenyl)}}{\overset{H}{C}} \end{array}\right]_x$ linear
addition

B. 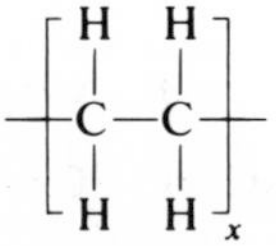 $\left[\begin{array}{c} \underset{H}{\overset{H}{C}}\text{—}\underset{H}{\overset{H}{C}} \end{array}\right]_x$ branched, addition
occasional C's have —CH₃
or —C₂H₅ branches

C. 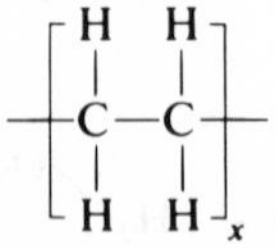 $\left[\begin{array}{c} \underset{H}{\overset{H}{C}}\text{—}\underset{H}{\overset{H}{C}} \end{array}\right]_x$ linear
addition

D. 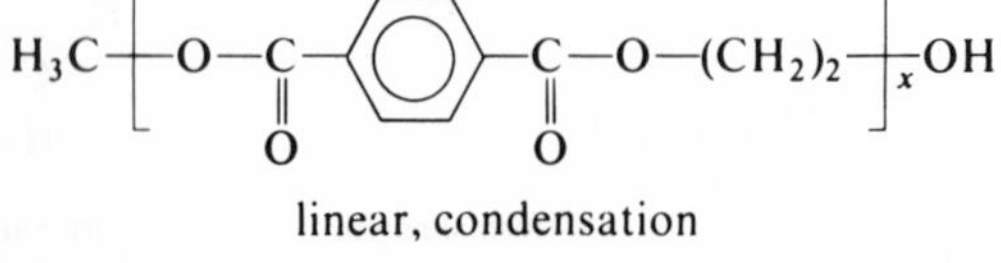 $H_3C\left[O\text{—}\underset{O}{\overset{}{C}}\text{—}\langle\text{benzene}\rangle\text{—}\underset{O}{\overset{}{C}}\text{—}O\text{—}(CH_2)_2\right]_x OH$ (CH₃OH
split out)

linear, condensation

E. $H\!-\!\Big[N\!-\!(CH_2)_6\!-\!N\!-\!C\!-\!(CH_2)_4\!-\!C\Big]_x\!\!-\!OH$ (H₂O out) linear, condensation

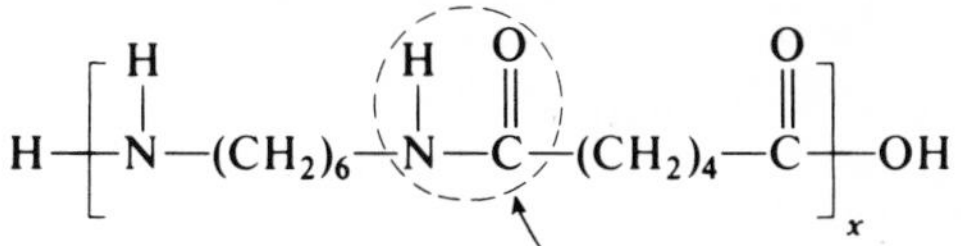

characteristic nylon linkage

F. 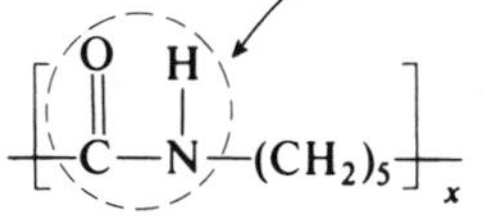 $\Big[C\!-\!N\!-\!(CH_2)_5\Big]_x$ (nothing out) monomer made

from $HO\!-\!C\!-\!(CH_2)_5\!-\!N\!-\!H$

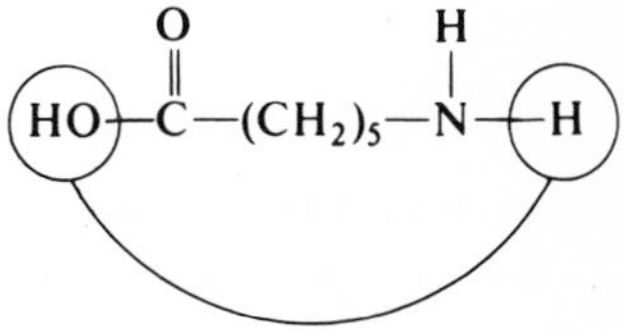

linear, condensation

G. 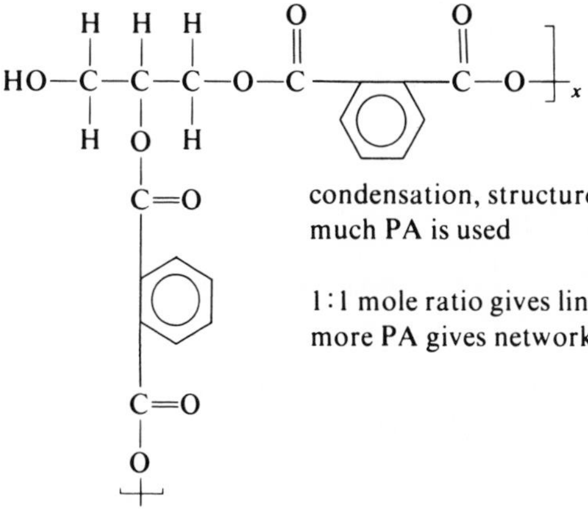 $HO\!-\!C\!-\!C\!-\!C\!-\!O\!-\!C\!-\!\bigcirc\!-\!C\!-\!O\Big]_x$

condensation, structure depends on how much PA is used

1:1 mole ratio gives linear or branched, more PA gives network

H. 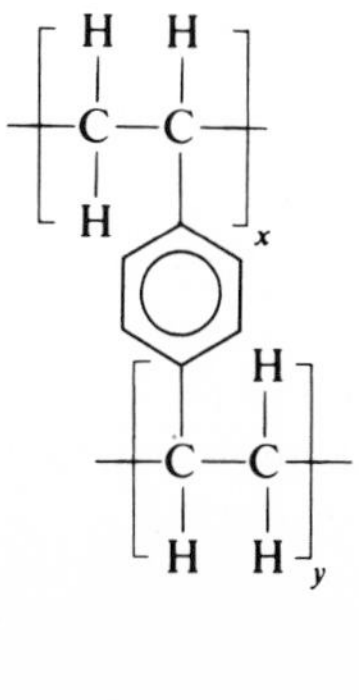 $\Big[C\!-\!C\Big]_x$ network addition

$\Big[C\!-\!C\Big]_y$ (monomer is 4-functional)

I.

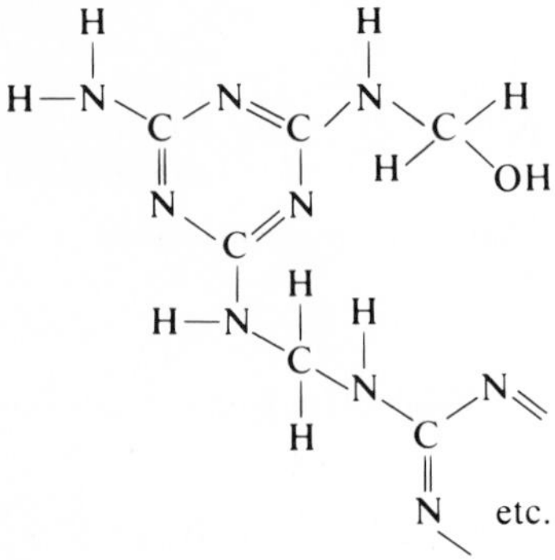

With more than 1 mole formaldehyde/mel.
get network, condensation

J.

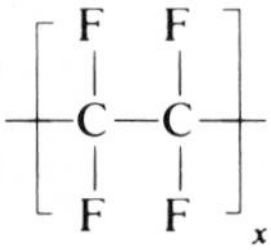

linear
addition

K.

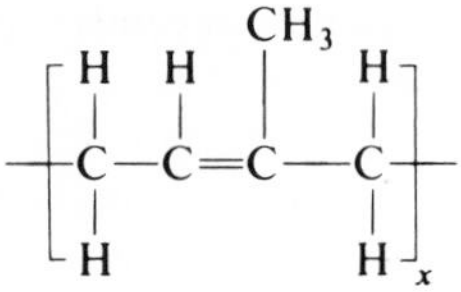

network, addition

natural rubber is largely 1,4, crosslinked with
sulfur through the double bonds in chain

L.

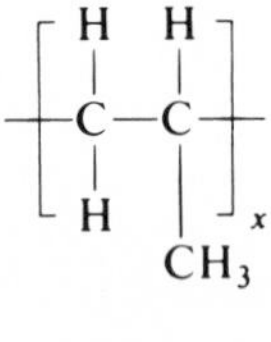

linear
addition

M.

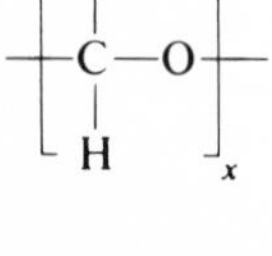

linear
addition

(only other double
bond which forms
addition polymers)

N. $H-\left[O-\bigcirc-\underset{\underset{CH_3}{|}}{\overset{\overset{CH_3}{|}}{C}}-\bigcirc-O-\underset{\underset{O}{\|}}{C}\right]_x Cl$ (HCl out) linear, condensation

O. $\left[-O-\bigcirc-\underset{\underset{CH_3}{|}}{\overset{\overset{CH_3}{|}}{C}}-\bigcirc-O-\underset{\underset{H}{|}}{\overset{\overset{H}{|}}{C}}-\underset{\underset{O}{|}}{\overset{\overset{H}{|}}{C}}-\underset{\underset{H}{|}}{\overset{\overset{H}{|}}{C}}-\right]_x$
                                                                    |
                                                                    H

Note: get both condensation —OH + Cl → HCl
and ring scission

$$-OH + H_2-\underset{\underset{O}{\diagdown\diagup}}{\overset{\overset{H}{|}}{C}}-\underset{\underset{H}{|}}{\overset{\overset{H}{|}}{C}}-C- \rightarrow -O-\underset{\underset{H}{|}}{\overset{\overset{H}{|}}{C}}-\underset{\underset{H}{|}}{\overset{\overset{O-H}{|}}{C}}-\underset{\underset{H}{|}}{\overset{\overset{H}{|}}{C}}-$$

linear, condensation
for $x < 8$, liquid epoxy ⎫ generally crosslinked later with
    $8 < x < 20$, solid epoxy ⎬ diamines or acid anhydrides
                             ⎭ thru —OH and terminal

$-\underset{\underset{O}{\diagdown\diagup}}{\overset{\overset{H}{|}}{C}}-CH_2$ groups

$x \simeq 100$, linear, "Phenoxy" plastic

P. $\left[-\underset{\underset{CH_3}{|}}{\overset{\overset{CH_3}{|}}{Si}}-O-\right]_x$ (HCl out)
                 linear, condensation

Note: commercial materials
contain some

$Cl-\underset{\underset{Cl}{|}}{\overset{\overset{CH_3}{|}}{Si}}-Cl$ which is 3-functional
and allows crosslinking.
Also, acetate may be
substituted for —Cl

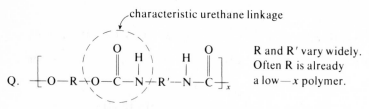

characteristic urethane linkage

Q. $\left[-O-R-O-C-N-R'-N-C- \right]_x$

R and R' vary widely.
Often R is already
a low—x polymer.

linear, condensation—nothing out,
but monomer can be considered

$-2H_2O$

can be crosslinked through amine groups,
or by using higher functional alcohols.

3
Bonding in Polymers

3.1 TYPES OF BONDS

Various types of bonding hold together the atoms in polymeric materials, as opposed to metals, for example, where only one type of bonding exists. These types are: (1) primary covalent, (2) hydrogen bonding, (3) dipole interaction, (4) van der Waals, and (5) ionic, examples of which are shown in Figure 3.1 (*1*). Hydrogen bonding, dipole interaction, van der Waals and ionic bonding are known collectively as secondary forces. The distinctions are not always clear cut; i.e., hydrogen bonds may be considered as the extreme of dipole interactions.

3.2 BOND DISTANCES AND STRENGTHS

Regardless of the type of bonding, the potential energy of the interacting atoms as a function of the separation between them is represented qualitatively by the potential function sketched in Figure 3.2. As the interacting centers are brought together from large separation, an increasingly great attraction tends to draw them together (negative potential energy). Beyond the separation r_m, as the atoms are brought closer together, their electronic 'atmospheres' begin to interact, and a powerful repulsion is set up. At r_m, the system is at a potential energy minimum—i.e., at its most probable or equilibrium separation, with r_m being the equilibrium bond distance. The depth of the potential well, ϵ, is the energy required to break the bond and separate the atoms completely.

Table 3.1 lists the approximate bond strengths and interatomic distances of the bonds encountered in polymeric materials. The important fact to notice here is how much stronger

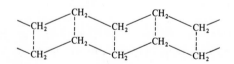

Figure 3.1.   Bonding in polymer systems (*1*).

the primary covalent bonds are than the others. As the material's temperature is raised and its thermal energy (kT) is thereby increased, the primary covalent bonds will be the last

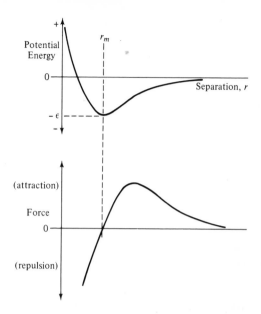

Figure 3.2.   Interatomic potential energy and force.

to dissociate, when the available thermal energy exceeds the dissociation energy.

TABLE 3.1   Bond Parameters ($1, 2$)

Bond Type	Interatomic Distance, r_m	Dissociation Energy, $-\epsilon$
Primary covalent	1–2 Å	50–200 Kcal/mole
Hydrogen bond	2–3	3–7
Dipole interaction	2–3	1.5–3
van der Waals	3–5	0.5–2
Ionic	2–3	10–20

3.3 BONDING AND RESPONSE TO TEMPERATURE

Now, in linear and branched polymers, only the secondary bonds hold the individual polymer chains together (neglecting temporary, purely mechanical entanglements). Thus, as the temperature is raised, a point will be reached where the forces holding the chains together become insignificant, and they are then free to slide past one another—i.e., to *flow* upon the application of stress. Therefore, *linear* and *branched* polymers are generally *thermoplastic*. The main chains in a crosslinked polymer, on the other hand, are held together by the same primary covalent bonds as are the atoms in the main chains. When the thermal energy exceeds the dissociation energy of the primary covalent bonds, both main chain and crosslink bonds fail randomly, and the polymer degrades. Hence, *crosslinked polymers are thermosetting*.

There are some exceptions to these generalizations. It is occasionally possible for secondary forces to make up for in quantity what they lack in quality. Polyacrylonitrile

$$\left(\left[\begin{array}{c} \overset{\text{H}}{\underset{\text{H}}{\overset{|}{\underset{|}{C}}}} - \overset{\text{H}}{\underset{\underset{\substack{\|\| \\ N}}{C}}{\overset{|}{\underset{|}{C}}}} \end{array}\right]_x\right), \text{ for example, is capable of strong interactions}$$

at every other carbon atom along the chains. If these secondary forces could be broken one by one (i.e., unzipped), it would behave as a typical thermoplastic. This is impossible, of course, and, by the time enough of the secondary bonds have been dissociated to free the chains and allow flow, the dissociation energy of the carbon-carbon main-chain bonds has been exceeded and the material degraded. Extreme stiffness of the polymer chain also contributes to this sort of behavior. For these reasons, cellulose, which, although linear, is both strongly hydrogen bonded and stiff, behaves as a thermoset. Polytetrafluoroethylene (Teflon TFE) is another example, due to the close packing and extensive secondary bonding of the main chains.

3.4 ACTION OF SOLVENTS

The action of solvents on polymers is in many ways similar to that of heat. Appropriate solvents (i.e., those which can form secondary bonds to the polymer chains) can penetrate, replace the interchain secondary bonds, and thereby pull apart and dissolve linear and branched polymers. The polymer-solvent secondary bonds cannot overcome primary valence crosslinks, however, so crosslinked polymers are not soluble, although they may swell extensively. (Try soaking a rubber band in benzene overnight.) The amount of swelling is, in fact, a convenient measure of the extent of crosslinking. A lightly crosslinked polymer (e.g. a rubber band) will swell tremendously, while one with extensive crosslinking (e.g., an ebonite—hard rubber—bowling ball) will not swell noticeably at all.

REFERENCES

1. Platzer, N. Progress in polymer engineering. *Ind. and Eng. Chem.*, **61,** no. 5, p. 10, 1969.

2. Miller, M. L. *The Structure of Polymers.* Reinhold Publishing Corp., New York, 1966.

4
Stereoisomerism

It is obvious now that the chemical nature of a polymer is of considerable importance in determining the polymer's properties. Of comparable significance is the way the molecules are arranged within the individual polymer chain.

4.1 INTRODUCTION

The carbon atom is normally (exclusively, for our purposes) tetravalent. In compounds such as methane (CH_4) and carbon tetrachloride (CCl_4), the four identical substituents surround it in a symmetrical tetrahedral geometry. If the substituent atoms are not identical, the symmetry is destroyed, but the general tetrahedral pattern is maintained. This is still true for each carbon atom in the interior of a linear polymer chain, where two of the substituents are chains. If a polyethylene chain were to be stretched out, for example, the carbon atoms in the chain backbone would lie in a zig-zag fashion in a plane, with the hydrogen substituents on either side of the plane (Figure 4.1). In the case of polyethylene, with a perfectly symmetrical repeating unit, this structural arrangement is mostly of academic importance. With polymers in which the repeating unit is not symmetrical, however, it assumes great importance.

4.2 STEREOISOMERISM IN VINYL POLYMERS

Before beginning a discussion of structural arrangements in vinyl polymers, it must be pointed out that they polymerize

Figure 4.1.   The geometry of a polyethylene chain.

almost exclusively in a head-to-tail configuration. The reasons for this are the steric interference of similar substituent groups and the electrostatic repulsion of groups with similar polarities. There are then three possible ways in which the unsymmetrical group may be arranged with respect to the carbon-carbon backbone plane (see Figure 4.2)—i.e., three *stereoisomers*.

Atactic   A random arrangement of the unsymmetrical groups is known as an *atactic* structure. Its lack of regularity has important consequences.

Isotactic   The structure in which all of the groups are lined up on the same side of the backbone plane is termed *isotactic*.

Syndiotactic   Alternating placement of the group on either side of the chain is the *syndiotactic* structure.

The terms above were coined by Dr. Guilio Natta, who shared the 1964 Nobel Chemistry Prize for his work in the area.

Although useful for descriptive purposes, the planar zig-zag arrangement of the main-chain carbon atoms is not always the

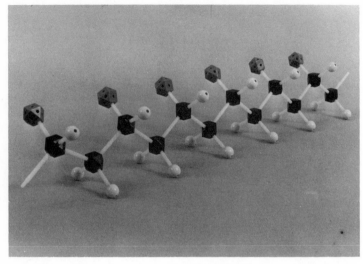

Figure 4.2a.   Stereoisomerism in vinyl polymers—isotactic.

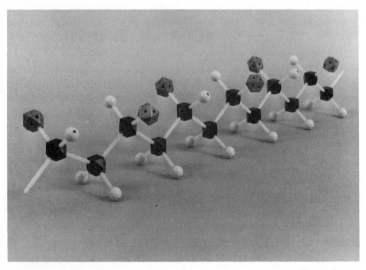

Figure 4.2b.   Stereoisomerism in vinyl polymers—syndiotactic.

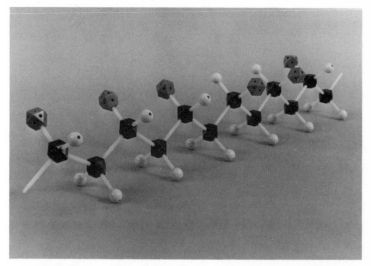

Figure 4.2c. Stereoisomerism in vinyl polymers—atactic.

one preferred by nature; i.e., it is not necessarily the minimum free-energy configuration. In the case of isotactic and syndiotactic polypropylene, for example, the preferred (minimum-energy) configurations are quite regular, while the atactic is irregular (Figure 4.3) (*1*). Atactic polypropylene has a consistency somewhat like used chewing gum, whereas the stereoregular forms are hard, rigid plastics.

The type of stereoregularity described above is a direct result of the dissymmetry of vinyl monomers. It is established in the polymerization reaction, and no amount of twisting and turning the chain about its bonds can convert one stereoisomer into another (molecular models are a real help, here).

Example 1. Both isotactic and atactic polymers of propylene oxide, $H-\overset{\displaystyle H}{\underset{\displaystyle \diagdown O \diagup}{C}}-\overset{\displaystyle H}{C}-CH_3$, have been prepared by ring scission polymerization.

a. Write the general structural formula for the polymer.
b. Indicate how the atactic, isotactic and syndiotactic structures differ.

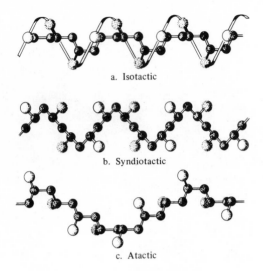

a. Isotactic

b. Syndiotactic

c. Atactic

Figure 4.3.   Configurations of polypropylene chains. The large balls
represent methyl (—CH$_3$) groups, and hydrogen atoms
are not shown (*1*). (Copyright 1961 by Scientific American, Inc. All rights reserved.)

Solution.

a. $\left[\text{O} - \underset{\underset{\text{H}}{|}}{\overset{\overset{\text{H}}{|}}{\text{C}}} - \underset{\underset{\text{CH}_3}{|}}{\overset{\overset{\text{H}}{|}}{\text{C}}} \right]_x$

b. —O—C—C—O—C—C—O—C—C—   atactic

—O—C—C—O—C—C—O—C—C—   isotactic

$$-O-\underset{\underset{\displaystyle H}{|}}{\overset{\overset{\displaystyle H}{|}}{C}}-\underset{\underset{\displaystyle CH_3}{|}}{\overset{\overset{\displaystyle H}{|}}{C}}-O-\underset{\underset{\displaystyle H}{|}}{\overset{\overset{\displaystyle H}{|}}{C}}-\underset{\underset{\displaystyle H}{|}}{\overset{\overset{\displaystyle CH_3}{|}}{C}}-O-\underset{\underset{\displaystyle H}{|}}{\overset{\overset{\displaystyle H}{|}}{C}}-\underset{\underset{\displaystyle CH_3}{|}}{\overset{\overset{\displaystyle H}{|}}{C}}- \qquad \text{syndiotactic}$$

4.3 STEREOISOMERISM IN DIENE POLYMERS

Another type of stereoisomerism arises in the case of poly 1,4-dienes because of the impossibility of rotation about a double bond. The substituent groups on the double-bonded carbons may either be on the same side of the chain (*cis*) or on opposite sides (*trans*), as shown in Figure 4.4 for 1,4-polyisoprene. The chains of cis 1,4-polyisoprene assume a tortured, irregular configuration because of the steric interference of the substituents adjacent to the double bonds. This stereoisomer is familiar as natural rubber. Trans 1,4-polyisoprene chains assume a regular structure. The polymer is gutta-percha, a hard, tough material, long used as a golf-ball cover.

Note that stereoisomerism in poly 1,4-dienes does not depend on dissymmetry of the repeating unit. The same isomers

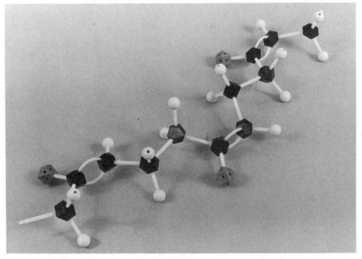

Figure 4.4a.   Stereoisomers of 1,4-polyisoprene—cis.

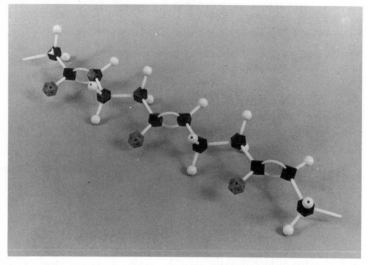

Figure 4.4b.   Stereoisomers of 1,4 polyisoprene—trans.

are possible with butadiene, in which all carbon substituents are hydrogen.

Example 2. Identify all the possible structural and stereo-isomers which can result from the addition polymerization of butadiene,

$$
\begin{array}{cccc}
H & & H \\
| & & | \\
C\!=\!C\!-\!C\!=\!C \\
| & | & | & | \\
H & H & H & H
\end{array}
$$

Solution. It will be recalled from Chapter 2 that butadiene may undergo 1,4 addition or 1,2 addition. The 1,4 polymer can have cis and trans isomers like the 1,4 polyisoprene in Figure 4.4. The 1,2 polymer,

$$
\begin{array}{c}
\begin{array}{cc}
H & H \\
| & | \\
\end{array} \\
\left[\!\!\begin{array}{cc}
C & -\!C \\
| & | \\
H & C\!-\!H \\
& \| \\
\end{array}\!\!\right]_x \\
H\!-\!C\!-\!H
\end{array}
$$

can have atactic, isotactic and syndiotactic stereoisomers like any vinyl polymer.

Example 3. Identify all the possible structural and stereoisomers which can result from the addition polymerization of

$$\underset{\displaystyle \overset{|}{H} \quad\quad \overset{|}{Cl} \quad \overset{|}{H} \quad \overset{|}{H}}{\overset{\displaystyle \overset{|}{H} \quad\quad\quad\quad \overset{|}{H}}{C=C-C=C}}$$

chloroprene, $\underset{H\ \ Cl\ \ H\ \ H}{\overset{H\quad\quad H}{C=C-C=C}}$ (the commercial polymer of

which is known as Neoprene).

Solution. As in Example 2, the 1,4 polymer can have cis and trans stereoisomers. This unsymmetrical diene monomer can also undergo both 1,2 addition

$$\begin{bmatrix} & H & Cl \\ & | & | \\ -C & - & C- \\ & | & | \\ & H & C-H \\ & & \| \\ & & H-C-H \end{bmatrix}_x$$

and 3,4 addition

$$\begin{bmatrix} H & H \\ | & | \\ -C & - & C- \\ | & | \\ Cl-C & H \\ \| \\ H-C-H \end{bmatrix}_x$$

. *Each* of these *structural* isomers may have

atactic, isotactic, or syndiotactic stereoisomers. Thus, in principle, at least, there are *eight* different isomers of polychloroprene possible.

REFERENCE

1. Natta, G. Precisely Constructed Polymers. *Sci. Amer.*, **205**, no. 2, p. 33, 1961.

5
Crystallinity in Polymers

It was pointed out in the previous chapter that the stereo-specific arrangement of the atoms in a polymer chain exerted a significant influence on the properties of the bulk polymer. To appreciate why this is so, the subject of polymer crystallinity is introduced here.

5.1 REQUIREMENTS FOR CRYSTALLINITY

Although the precise nature of crystallinity in polymers is still under investigation, a number of facts have long been known about the requirements for polymer crystallinity. First, an ordered, regular chain structure is necessary to allow the chains to pack into a regular crystal lattice. Thus, stereo-regular polymers are more likely to be crystalline than those which have irregular chain structures. Irregular, protruding side chains interfere with the arrangement of the main chains in a regular lattice and hinder crystallinity. Second, no matter how regular the chains, the secondary forces holding the chains together in the crystal lattice must be strong enough to overcome the disordering effect of thermal energy, so hydrogen bonding or strong dipole interactions promote crystallinity, and, other things being equal, raise the crystalline melting point.

X-ray studies show that there are numerous polymers which do not meet the above criteria and show no traces of crystallinity, i.e.; they are completely *amorphous*. On the other hand, despite intensive efforts, no one has succeeded in producing a completely crystalline polymer. The crystalline content may in certain cases be pushed up to on the order of 98 percent, but there always remains a few per cent of amorphous material.

In the case of metals, defect concentrations are on the order of parts per million, so they can be considered perfectly crystalline in comparison with polymers. The fiber industry takes advantage of the fact that the degree of crystallinity of certain polymers can be increased by a drawing operation—i.e., by stretching the fibers. The increasing crystallinity can be followed by several techniques.

5.2 THE FRINGED MICELLE MODEL

The first attempt to explain the observed properties of crystalline (the word should be prefaced by semi-, but rarely is) polymers was the *fringed micelle* model (Figure 5.1). This

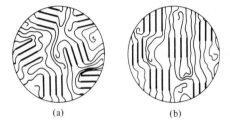

(a)   (b)

Figure 5.1.   The fringed micelle model. (a) unoriented, (b) chains oriented by applied stresses.

model pictures crystalline regions known as fringed micelles or *crystallites* interspersed in an amorphous matrix. The crystallites, whose dimensions are on the order of several hundred Angstroms, are small volumes in which the chains are regularly aligned parallel to one another, tightly packed into a crystal lattice. The chains, however, are many times longer than the dimensions of an individual crystallite, so they pass from a crystalline area through an amorphous area, back into another crystallite, and so on. This model nicely explains the coexistence of crystalline and amorphous areas in polymers, and it also explains the increase in crystallinity observed in the drawing operation. Stretching the polymer orients the chains in the direction of the stress, increasing the alignment in the amorphous areas and producing greater degrees of crystallinity (Figure 5.1b).   Since the chains pass at random from one

crystallite to another, it is easily seen why perfect crystallinity can never be achieved. It also explains why the effects
of crystallinity on mechanical properties are in many ways
similar to those of crosslinking, because, like crosslinks, the
crystallites tie the individual chains together. Unlike crosslinks, though, the crystallites will generally melt before the
polymer degrades, and solvents which form extremely powerful
secondary bonds with the chains can dissolve them.

The fringed micelle model, although now largely superseded
by more recent developments, still does a fine job of predicting
the effects of crystallinity on mechanical properties of polymers—much better than many of the newer concepts which, at
this stage of the game, still require a bit of hand-waving to
square with the observed facts.

5.3 FOLDED-CHAIN CRYSTALLITES (1, 2)

The first direct observations of the nature of polymer crystallinity resulted from the growth of single crystals from dilute
solution. Either by cooling or evaporation of solvent, pyramidal or plate-like polymer crystals (lamellae) were precipitated from dilute solutions (Figure 5.2). These crystals were
several hundred thousand Angstroms along a side and about
100Å thick. This was fine, except that x-ray measurements

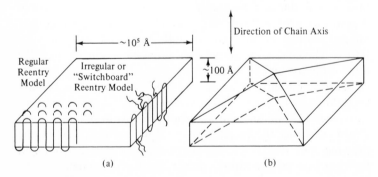

Figure 5.2.   Polymer single crystals. (a) flat lamellae, (b) pyramidal
lamellae. Two concepts of chain reentry are illustrated.

showed that the polymer chains were aligned perpendicular to the large flat faces of the crystals, and it was known that the individual chains were on the order of 1000Å long. How could a chain fit into a crystal one-tenth its length? The only answer is that the chain must fold back on itself, as shown in Figure 5.2.

This folded-chain model has been well substantiated for single polymer crystals. The lamellae are about 50–60 carbon atoms thick, with about five carbon atoms in a direct reentry fold. These atoms in the fold, of course, can never be part of a crystal lattice. Just because such folded-chain lamellae are obtained in single crystals grown from dilute solution is no guarantee that similar structures exist in bulk polymer samples, but evidence is piling up that they do. Figure 5.3 illustrates a

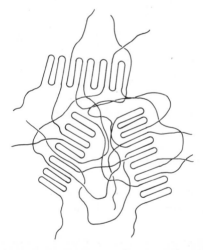

Figure 5.3. Compromise model showing lamellae tied together by interlamellar amorphous chains.

model combining the folded-chain lamellae with the interlamellar amorphous material tying the lamellae together in a bulk polymer. Additional orientation and crystallization in the interlamellar amorphous regions, as in the fringed-micelle model, is usually invoked to explain the increase in crystallinity with drawing.

5.4 SPHERULITES

Not only are polymer chains often arranged to form crystallites, but these crystallites are often arranged in larger aggregates known as *spherulites*. These spherulites grow radially from a point of nucleation until other spherulites are encountered. Thus, the size of the individual spherulites can be controlled by the number of nucleii present, more nucleii resulting in more but smaller spherulites. Spherulites are in some ways similar to the grain structure in metals. They are typically about 0.01 mm in diameter and have a Maltese Cross appearance between crossed polaroids. Figure 5.4 shows how the

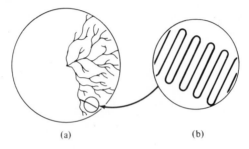

(a)   (b)

Figure 5.4.   Structure of spherulites.   (a) branching of lamellae, (b) orientation of chains in lamellae.

polymer chains are thought to be arranged in the spherulites. Large spherulites contribute to brittleness in polymers. To avoid this, nucleating agents are often added or the polymer is shock cooled to promote smaller spherulites.

5.5 THE EFFECTS OF CRYSTALLINITY ON POLYMER PROPERTIES

The presence of crystalline material in polymers strongly influences their properties. This influence is most profound on the mechanical properties.

Since the polymer chains are packed together more efficiently and tightly in the crystalline areas than the amorphous, the crystallites will a higher density. Thus low-density (0.915 g/cc) polyethylene is about 60 percent crystalline, while high-density (0.97 g/cc) polyethylene is about 95 percent crystalline. (To confuse the issue, low-density polyethylene is usually made by a high-pressure process and high-density by a low-pressure process. Both descriptive terms are used.) Density is, in fact, a convenient measure of the degree of crystallinity.

In the case of polyethylene, the difference in density arises mainly from short (2–6 carbon atoms) branching, a result of side reactions during polymerization. These occasional branches prevent packing into a crystal lattice in their immediate vicinity and thus lower the degree of crystallinity.

Since the polymer chains are more tightly packed in the crystalline areas than in the amorphous areas, there are more of them available per unit area to support a stress. Also, since they are in close and regular contact over relatively long distances in the crystallites, the net forces holding them together are far greater than in the amorphous regions. Thus, crystallinity can significantly increase the strength and rigidity of a polymer. For this reason, the stereoregular polypropylenes, which can and do crystallize, are hard and rigid plastics, while the irregular atactic polymer is amorphous and soft and sticky. The influence of the degree of crystallinity on some of the properties of polyethylene is illustrated in Table 1.

The optical properties of polymers also depend on crystallinity. When light passes between two phases with different refractive indices, some of it is scattered at the interface (e.g., a large salt crystal is transparent, but table salt appears white because light must pass alternately from air to salt many times). Crystalline polymers are actually two-phase systems, with a crystalline phase dispersed in an amorphous matrix. The denser crystalline areas have a higher refractive index than the amorphous ones, so crystalline polymers are either opaque or translucent. As the dimensions of the crystallites become appreciably smaller than the wavelength of light, the amount of light scattered decreases, and a polymer with very small crystallites and a low degree of crystallinity might appear transparent, but there is still some question as to whether such materials really exist. So, in general, *transparent polymers are completely amorphous.* The corollary is not necessarily true; i.e., lack of transparency in a polymer may be due to crystallinity, but it can also be caused by fillers, etc. If the polymer is known to

TABLE 1   The Influence of Crystallinity on Some of the
Properties of Polyethylene[a]

Commercial Product	Low Density	Medium Density	High Density
density range, g/cc	0.910–0.925	0.926–0.940	0.941–0.965
approximate % crystallinity	60–70	70–80	80–95
branching, equivalent CH_3 groups/1000 C atoms	15–30	5–15	1–5
crystalline melting point, °C	110–120	120–130	130–136
hardness, Shore D	41–46	50–60	60–70
tensile modulus, psi	0.14–0.38 × 10⁵	0.25–0.55 × 10⁵	0.6–1.8 × 10⁵
tensile strength, psi	600–2300	1200–3500	3100–5500
flexural modulus, psi	0.08–0.6 × 10⁵	0.6–1.15 × 10⁵	1.0–2.6 × 10⁵

[a]It must be kept in mind that mechanical properties are influenced by factors other than the degree of crystallinity (molecular weight, in particular).

be a pure homopolymer, however, translucency is a sure sign of crystallinity. In principle, a completely crystalline polymer would be transparent, but there can be no such thing.

Example 1. Explain the following facts:

a. Poly(ethylene) and poly(propylene) produced with stereo-specific catalysts are each fairly rigid, translucent plastics, while a 65-35 copolymer of the two, produced in exactly the same manner, is a soft, transparent rubber.

b. A plastic is commercially available which is similar in appearance and mechanical properties to the poly(ethylene) and poly(propylene) described in (a) but which consists of 65 percent ethylene and 35 percent propylene units. The two components of this plastic cannot be separated by any physical or chemical means without degrading the polymer. *Solution.*

a. The polyethylene produced with these catalysts is linear and thus highly crystalline. The polypropylene is isotactic and also highly crystalline. The crystallinity confers mechanical strength and translucency. The 65-35 copolymer, ethylene-propylene rubber (EPR) is a *random* copolymer, so the CH_3 groups from the propylene monomer are arranged at irregular intervals along the chain, preventing packing in a regular crystal lattice, giving an amorphous, rubbery polymer.

b. Since the components cannot be separated, they must be chemically bound within the chains. The properties indicate a crystalline polymer, so the CH_3 groups from the propylene cannot be spaced irregularly as in the random copolymer in (a). Thus, these materials must be *block* copolymers of ethylene and stereoregular polypropylene. The long blocks of ethylene pack into a polyethylene lattice, and the propylene blocks pack into a polypropylene lattice.

Example 2. Polyvinyl alcohol is made by the hydrolysis of polyvinyl acetate because the vinyl alcohol monomer is unstable.

Polyvinyl acetate   Polyvinyl alcohol   Acetic acid

$$
\left[\begin{array}{c} H \quad H \\ | \quad | \\ -C-C- \\ | \quad | \\ H \quad O \\ | \\ C=O \\ | \\ CH_3 \end{array}\right]_x + H_2O \rightarrow \left[\begin{array}{c} H \quad H \\ | \quad | \\ -C-C- \\ | \quad | \\ H \quad OH \end{array}\right]_x + H_3C-\overset{\displaystyle O}{\overset{\|}{C}}-OH
$$

The extent of reaction may be controlled to yield polymers with anywhere from 0–100 percent of the original acetate groups hydrolyzed. Pure polyvinyl acetate (0 percent hydrolyzed) is insoluble in water. It has been observed, however, that, as the extent of hydrolysis is increased, the polymers become more water soluble, up to about 87 percent hydrolysis, after which further hydrolysis *decreases* water solubility at room temperature. Explain briefly.

Solution. The normal polyvinyl acetate is atactic, and the irregular arrangement of the acetate side groups renders it completely amorphous. Water cannot form strong enough secondary bonds with the chains to dissolve it. As the acetate groups are replaced by hydroxyls, sites are introduced which can form strong hydrogen bonds with water, thereby increasing the solubility. At high degrees of substitution, replacement of the bulky acetate groups with the compact hydroxyls allows the chains to pack into a crystal lattice. The hydroxyl groups provide hydrogen bonding sites between the chains which help hold them in the lattice, and thus solubility is reduced.

Example 3. Explain the following experiment: A weight is tied to the end of a polyvinyl alcohol fiber. The weight and part of the fiber are dunked in a beaker of boiling water. As long as the weight remains suspended, the situation is stable, but, when the weight is rested on the bottom of the beaker, the fiber dissolves.

Solution. As long as the weight is suspended from the fiber, the stress maintains the alignment of the chains in a crystal lattice, against the disordering effects of thermal energy and solvent (water) penetration. When the weight is rested on the bottom, the stress is removed, allowing the polymer to dissolve.

Example 4. When a rubber band is rapidly stretched, if placed to the lips, it can be felt to have warmed. If held in the stretched configuration long enough to reach room temperature again, and then suddenly released, it will cool perceptibly. Explain.

Solution. If stretched rapidly enough, the process will be adiabatic. Natural rubber, when stretched to high elongations, crystallizes as a result of chain alignment. As in other materials, crystallization is an exothermic process, so the energy given off warms the material. The reverse is observed when the band is released and the crystals melt.

Example 5. In example 3 in chapter 2, polymers B, C, E, F, J, L and M are at least partially crystalline. Briefly discuss the important reason(s) for this in each case.

Solution. Polymers B and C. Polyethylene does not have any bulk side groups (as long as branching is not excessive) and can therefore pack into a lattice held together by Van der Waals forces.

Polymers E and F. Nylons lack bulky side groups and, in addition, can form strong hydrogen bonds between chains, giving higher crystalline melting points than polyethylene (6/6 = 264C, 6 = 215C).

Polymer J. The reasons for crystallinity in polytetrafluoroethylene are the same as those in polyethylene.

Polymer L. The stereoregular forms of polypropylene can

pack into a regular lattice, held together by Van der Waals forces. The atactic form is amorphous.

Polymer M. As with polyethylene, polyformaldehyde has no bulky side groups to hinder packing. In addition, the ether linkage provides dipole interaction sites which hold the chains together more strongly than the Van der Waals forces in polyethylene, giving a higher crystalline melting point (175C).

REFERENCES

1. Geil, P. H. *Polymer Single Crystals.* John Wiley & Sons, Inc., New York, 1963.

2. Oppenlander, G. C. Structure and properties of crystalline polymers. *Sci.*, **159,** no. 1311, 1968.

6
Characterization of Molecular Weight

6.1 INTRODUCTION

With the exception of a few naturally occurring polymers, all polymers consist of molecules with a *distribution* of chain lengths. It is therefore necessary to characterize the entire distribution quantitatively or at least to define and measure average chain lengths or molecular weights for these materials, as many important properties of the polymer depend on these quantities.

The concept of an *average* molecular weight causes some initial difficulty because we're used to thinking in terms of ordinary low molecular weight compounds in which the molecules are identical; there is a single, well-defined molecular weight for the compound. Where the molecules in a sample vary in size, however, the results depend on how you count.

In the case of pure, low molecular weight compounds, the molecular weight is defined as

$$M = W/N \qquad (6\text{-}1)$$

where W = total sample weight
and N = number of moles in the sample.

6.2 AVERAGE MOLECULAR WEIGHTS

Where a distribution of molecular weights exists, a *number-average* molecular weight, $\overline{M}_n$, may be defined in an analogous fashion to (6-1)

44

$$\overline{M}_n = \frac{W}{N} = \frac{\sum_{x=1}^{\infty} n_x M_x}{\sum_{x=1}^{\infty} n_x} = \frac{n_1 M_1}{\Sigma n_x} + \frac{n_2 M_2}{\Sigma n_x} + \cdots = \sum_{x=1}^{\infty} \left(\frac{n_x}{N}\right) M_x$$

(6-2)

where W = total sample weight = $\sum_{x=1}^{\infty} n_x M_x$,

N = total *number of moles* in the sample (of all

sizes) = $\sum_{x=1}^{\infty} n_x$,

n_x = *number of moles* of x-mer,

M_x = molecular weight of x-mer,

and (n_x/N) = mole fraction of x-mer.

Any analytical technique which determines the *number* of moles present in a sample of known weight, regardless of their size, will give the number-average molecular weight.

Rather than count the number of molecules of each size present in a sample, it is possible to define an average in terms of the *weights* of molecules present at each size level. This is the weight-average molecular weight, $\overline{M}_w$. A good way to illustrate the differences between the two averages is to consider the analogy of a mixture of various-sized ball bearings rolling down a trough into which successively larger slots have been cut (Figure 6.1). The smallest ball bearings fall into a compartment beneath the first slot, the next larger size into a compartment beneath the second slot, and so on, with the last compartment holding the largest ball bearings. (We are assuming that there are a few discrete bearing sizes and that therefore each compartment holds bearings of a single diameter.) The subscript i serves to identify the compartments in order of increasing slot size. The number-average ball bearing diameter, $\overline{D}_n$, is obtained by *counting* the numbers of ball bearings in each compartment

$$\overline{D}_n = \frac{\Sigma n_i D_i}{\Sigma n_i}$$

where D_i = diameter of bearings in compartment i

and n_i = number of bearings in compartment i,

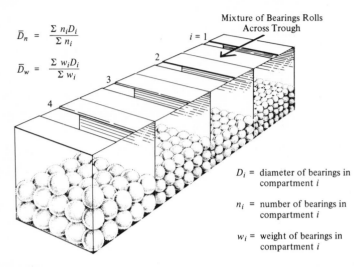

$$\bar{D}_n = \frac{\Sigma\, n_i D_i}{\Sigma\, n_i}$$

$$\bar{D}_w = \frac{\Sigma\, w_i D_i}{\Sigma\, w_i}$$

Mixture of Bearings Rolls
Across Trough

D_i = diameter of bearings in compartment i

n_i = number of bearings in compartment i

w_i = weight of bearings in compartment i

Figure 6.1.   Ball bearing analogy for average molecular weights.

and it is analogous to the number-average molecular weight. An equally valid average diameter, the *weight-average*, $\bar{D}_w$, is obtained by *weighing* the ball bearings in each compartment

$$\bar{D}_w = \frac{\Sigma\, w_i\, D_i}{\Sigma\, w_i}$$

where w_i = weight of all ball bearings in the i-th compartment
and D_i = diameter of the bearings in compartment i.

The analogous *weight-average*, $\bar{M}_w$, is then

$$\bar{M}_w = \frac{\Sigma\, w_x M_x}{\Sigma\, w_x} = \frac{w_1 M_1}{\Sigma\, w_x} + \frac{w_2 M_2}{\Sigma\, w_x} + \cdots = \sum \left(\frac{w_x}{W}\right) M_x = \frac{\Sigma\, n_x M_x^2}{\Sigma\, n_x M_x}$$

(6-3)

where w_x = weight of x-mer in sample = $n_x M_x$
and w_x/W = weight fraction of x-mer in sample.

Analytical procedures which, in effect, determine the weight of molecules at a given size level result in the weight-average molecular weight.

The number-average molecular weight is the *first moment* of the molecular weight distribution, analogous to the center of gravity (the first moment of the mass distribution) in mechanics. The weight-average molecular weight, the *second moment* of the distribution, corresponds to the radius of gyration in mechanics. Higher moments (e.g., $\overline{M}_z$, the third moment) may be defined and find occasional use.

It may be shown that $\overline{M}_w \geq \overline{M}_n$. The two are equal only for a *monodisperse* (all molecules the same size) polymer. The ratio $\overline{M}_w/\overline{M}_n$ is known as the *polydispersity index* and is a measure of the breadth of the molecular weight distribution. Values range from about 1.02 for carefully fractionated or anionic addition polymers to over 20 for many commercial polymers.

Example 1. Measurements on two essentially monodisperse fractions of a linear polymer, i and j, yield molecular weights of 10,000 and 40,000, respectively. Mixture (1) is prepared from one part by weight of i and two parts by weight of j. Mixture (2) contains two parts by weight of i and one of j. Determine the weight- and number-average molecular weights of mixtures (1) and (2).

Solution. For mixture (1),

$$n_i = \frac{1}{10,000} = 1 \times 10^{-4}$$

$$n_j = \frac{2}{40,000} = 0.5 \times 10^{-4}$$

$$\overline{M}_n = \frac{\Sigma n_x M_x}{\Sigma n_x} = \frac{(1 \times 10^{-4})(10^4) + (0.5 \times 10^{-4})(4 \times 10^4)}{1 \times 10^{-4} + 0.5 \times 10^{-4}}$$

$$= 2 \times 10^4$$

$$\overline{M}_w = \sum \left(\frac{w_x}{W}\right) M_x = \left(\frac{1}{3}\right)(1 \times 10^4) + \left(\frac{2}{3}\right)(4 \times 10^4)$$

$$= 3 \times 10^4.$$

For mixture (2),

$$n_i = \frac{2}{10,000} = 2 \times 10^{-4}$$

$$n_j = \frac{1}{40,000} = 0.25 \times 10^{-4}$$

$$\overline{M}_n = \frac{(2 \times 10^{-4})(10^4) + (0.25 \times 10^{-4})(10^4)}{2 \times 10^{-4} + 0.25 \times 10^{-4}} = 1.33 \times 10^4$$

$$\overline{M}_w = \left(\frac{2}{3}\right)(1 \times 10^4) + \frac{1}{3}(4 \times 10^{-4}) = 2 \times 10^4.$$

6.3 DETERMINATION OF AVERAGE MOLECULAR WEIGHTS

In general, techniques for the determination of average molecular weights fall into two categories: absolute and relative. In the former, measured quantities are theoretically related to the average molecular weight; in the latter, a quantity is measured which is in some way related to molecular weight, but the exact relation must be established by calibration with one of the absolute methods.

A. Absolute Methods

End-group analysis. If the chemical nature of the end groups on the polymer chains is known, standard analytical procedures may sometimes be employed to determine the concentration of the end groups and thereby of the polymer molecules, giving directly the number-average molecular weight. For example, in a linear polyester formed from a stoichiometrically equivalent batch, there is, on the average, one unreacted acid group and one unreacted —OH group per molecule. These groups may sometimes be analyzed by appropriate titration. If an addition polymer is known to terminate by disproportionation (see chapter 10), there will be one double bond per every two molecules which may be detectable quantitatively by halogenation or by infrared measurements. Other possibilities include the use of a radioactively tagged initiator which remains in the chain ends.

These methods have one drawback in addition to the necessity of knowing the nature of the end groups. As the molecular weight increases, the concentration of end groups (number per unit volume) decreases, and the measurement sensitivity drops off rapidly. For this reason, they are generally limited to the range $\overline{M}_n < 10,000$.

Colligative property measurements. When a solute is added to a solvent, it causes a change in the activity and chemical potential (partial molal Gibbs free energy) of the solvent. The magnitude of the change is directly related to the solute concentration. For example, when pure water is boiled, the chemical potentials of the liquid and the vapor in equilibrium with it are the same. If, now, some salt is added to the water, it lowers the chemical potential of the liquid water. In order to reestablish equilibrium with the pure water vapor above the salt solution, the temperature of the system must be raised, causing a boiling-point elevation. In a similar fashion, the addition of ethylene glycol antifreeze depresses the freezing point of water.

Freezing-point depression, boiling-point elevation, and a third technique, osmotic pressure, may be used to determine the number of moles of polymer per unit volume of solution and thereby establish the number-average molecular weight. The relevant thermodynamic equations for the three techniques are:

$$\lim_{c \to 0} \frac{\Delta T_b}{c} = \frac{RT^2}{\rho \, \Delta H_v \, \overline{M}_n} \quad \text{(boiling-point elevation)} \quad (6\text{-}4)$$

$$\lim_{c \to 0} \frac{\Delta T_f}{c} = \frac{-RT^2}{\rho \Delta H_f \, \overline{M}_n} \quad \text{(freezing-point depression)} \quad (6\text{-}5)$$

$$\lim_{c \to 0} \frac{\pi}{c} = \frac{RT}{\overline{M}_n} \quad \text{(osmotic pressure)} \quad (6\text{-}6)$$

where c = solute (polymer) concentration, weight/volume,

T = absolute temperature,

R = gas constant,

ΔH_v = solvent enthalpy of vaporization,

ΔH_f = solvent enthalpy of fusion,

ΔT_b = boiling-point elevation,

ΔT_f = freezing-point depression,

and π = osmotic pressure.

It is important to note that the thermodynamic equations apply only for ideal solutions, a condition which can only be reached in the limit of infinite dilution of the solute.

Freezing-point depression and boiling-point elevation require precise measurements of very small temperature differences. Although they are used occasionally, the difficulties involved have prevented their widespread application. Os-

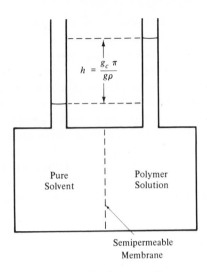

$$h = \frac{g_c \pi}{g\rho}$$

Pure
Solvent

Polymer
Solution

Semipermeable
Membrane

Figure 6.2.   Schematic diagram of an osmometer.

motic pressure, on the other hand, is the most common method of determining $\bar{M}_n$. A schematic diagram of an osmometer is shown in Fig. 6-2. The solution and solvent chambers are separated by a semipermeable membrane, one which ideally allows passage of solvent molecules but not solute molecules. The solvent flows through the membrane to dilute the solution. This is a natural consequence of the tendency of the system to increase its entropy, which is accomplished by the dilution of the solution. This dilution continues until the tendency toward further dilution is counterbalanced by the increased pressure in the solution chamber. At this point, the chemical potential of the solvent is the same in both chambers, and the pressure difference between the chambers is the osmotic pressure, π. By making measurements at several concentrations, plotting $\frac{\pi}{c}$ versus c and extrapolating to zero concentration, the number-average molecular weight is established through (6-6).

One of the major difficulties with membrane osmometry is finding suitable semipermeable membranes. Ordinary cellophane or modifications of it are commonly used. Unfortunately, if the membrane is sufficiently tight to prevent the passage of low molecular-weight chains, the rate of solvent passage is slow, and it takes a long time to reach equilibrium. In practice, all membranes allow some of the low-molecular weight polymer in a distribution to sneak through. Also, as the average molecular weight increases, π decreases, so the measurement precision decreases. These factors usually limit the applicability of the technique to $50,000 < \overline{M}_n < 1,000,000$.

Another colligative technique which is beginning to find widespread application is *vapor-pressure osmometry*. A drop of polymer solution and a drop of pure solvent are placed on separate thermistors within a closed chamber. Because of the solvent's lower chemical potential in the solution, solvent tends to diffuse from the pure drop to the solution drop. When it reaches the solution drop, it condenses, giving off its latent heat of vaporization which in turn heats the solution drop. Diffusion continues until the temperature of the solution drop reaches a point at which the chemical potential of solvent in each drop is the same. The equilibrium temperature difference is related directly to the molar solute (polymer) concentration and hence to the number-average molecular weight. Equilibrium is established in a few minutes, allowing rapid determination of $\overline{M}_n$.

The techniques discussed to this point establish the *number* of molecules present per unit volume of solution, regardless of their size or shape. Other methods measure quantities which are related to the average *mass* of the molecules in solution, thereby giving the weight-average molecular weight. One of the more common of these is *light scattering*, which is based on the fact that the intensity of light scattered by a polymer molecule is proportional, among other things, to the square of its mass. A light-scattering photometer measures the intensity of scattered light as a function of the scattering angle. Measurements are made at several concentrations. By a double extrapolation to zero angle and zero concentration (*Zimm plot*) and with a knowledge of the dependence of the solution refractive index versus concentration, $\overline{M}_w$ is established. This technique also provides information on the solvent-solute interaction and on the configuration of the polymer molecules in solution, since the quantitative nature of the scattering depends also on the size of the particles. Light scattering is generally applicable over the range $10,000 < \bar{M}_w < 10,000,000$.

The weight-average molecular weight can also be obtained with an ultracentrifuge, which distributes the molecules according to their mass in a centrifugal force field.

B. Relative Methods   The molecular-weight determination techniques discussed to this point allow direct calculation of the average molecular weight from experimentally measured quantities through known theoretical relations. Sometimes, however, these relations are not known, and, although something is measured which is known to be related to molecular weight, one of the absolute methods above must be used to calibrate the technique.

A case in point is solution viscosity. It has long been known that relatively small amounts of dissolved polymer could cause tremendous increases in viscosity, and it is logical to assume that, at a given concentration, the larger molecules will impede flow more and give the higher viscosity. A quantitative basis for the treatment of solution viscosity was provided by Einstein in 1920. For a suspension of *rigid, noninteracting spheres*, he developed the relation

$$\eta_{rel} = \frac{\eta}{\eta_s} = 1 + 2.5\phi \qquad (6\text{-}7)$$

where η = viscosity of the suspension,

η_s = viscosity of the solvent,

η_{rel} = *relative viscosity*,

and ϕ = volume fraction of spheres.

After rearranging,

$$\left(\frac{\eta}{\eta_s} - 1\right) = \eta_{sp} = 2.5\phi \qquad (6\text{-}8)$$

where η_{sp} = *specific viscosity*.

Einstein's equation has been amply verified for very dilute suspensions of rigid spheres, where the assumptions in the derivation are met.

Polymer molecules in solution are of course not rigid and are not necessarily spherical, and, at finite concentrations, they are bound to interact with one another. All these factors depend on the interaction between the solute and solvent molecules and, hence, upon the particular polymer-solvent system and the temperature. Recognizing these limitations, (6-8) may be rewritten as

$$\left(\frac{\eta}{\eta_s} - 1\right) = \eta_{sp} = Kc \qquad (6\text{-}9)$$

where K is a function of the size of the dissolved molecules, their shape, their rigidity, the interactions between them, and the proportionality between volume fraction and concentration. Ultimately, then, K depends on the particular polymer-solvent system, the temperature, and the size of the molecules in solution. If one divides both sides by c,

$$\left(\frac{\eta}{\eta_s} - 1\right)\bigg/ c = \eta_{red} = K \qquad (6\text{-}10)$$

where η_{red} = *reduced viscosity*.

The effects of polymer-polymer interactions on the reduced viscosity may be eliminated by extrapolation to zero concentration

$$\lim_{c \to 0} \left(\frac{\eta_{sp}}{c}\right) = [\eta] \qquad (6\text{-}11)$$

where $[\eta]$ = *intrinsic viscosity*.

The intrinsic viscosity, then, should be a function of the size of the polymer molecules in solution, the polymer-solvent system and the temperature. If measurements are made at constant temperature using a specified solvent for a particular polymer, it should be related to the polymer's molecular weight.

Huggins proposed a relation between reduced viscosity and concentration for dilute polymer solutions ($\eta_{rel} < 2$):

$$\frac{\eta_{sp}}{c} = [\eta] + k'[\eta]^2 c \quad \text{(Huggins equation).} \qquad (6\text{-}12a)$$

Interestingly enough, k' turns out to be approximately equal to 0.4 for a variety of polymer-solvent systems, providing a convenient means of estimating dilute solution viscosity versus concentration if the intrinsic viscosity is known. By expanding the natural logarithm in a power series, it may be shown that an equivalent form of the Huggins equation is

$$\eta_{inh} = \frac{\ln \eta_{rel}}{c} = [\eta] - k''[\eta]^2 c \qquad (6\text{-}12b)$$

where η_{inh} = inherent viscosity
$k'' = k' - 0.5$.

Plots of the reduced and inherent viscosities are linear with concentration, (at least below concentrations of about 0.5 grams/deciliter) as predicted by the Huggins equation, with a common intercept, the intrinsic viscosity (Figure 6.3). Excep-

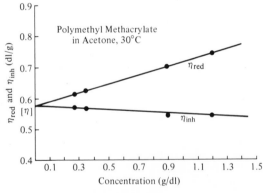

Figure 6.3. Determination of intrinsic viscosity (example 2).

tions occur with polyelectrolyte solutions, where the chemical nature of the polymer changes with concentration and the concept of intrinsic viscosity is not strictly applicable.

Note that the intrinsic viscosity has dimensions of reciprocal concentration. For some strange reason, concentrations are usually given in grams/deciliter (= 100 ml). In fact, the relative, specific, reduced, intrinsic and inherent viscosities are not true viscosities, and they don't have dimensions of viscosity. More appropriate terminology has been proposed but has not been widely adopted. Table 1 summarizes the various quantities defined and typical units.

TABLE 1 Solution Viscosity Terminology (1)

Quantity	Common Units	Common Name	Recommended Name
η	centipoise	solution viscosity	solution viscosity
η_s	centipoise	solvent viscosity	solvent viscosity
$\eta_{rel} = \dfrac{\eta}{\eta_s}$	dimensionless	relative viscosity	viscosity ratio
$\eta_{sp} = \dfrac{\eta - \eta_s}{\eta_s} = \eta_{rel} - 1$	dimensionless	specific viscosity	—
$\eta_{red} = \dfrac{\eta_{sp}}{c} = \dfrac{\eta_{rel} - 1}{c}$	$\dfrac{\text{deciliters}}{\text{gram}}$	reduced viscosity	viscosity number
$\eta_{inh} = \dfrac{\ln \eta_{rel}}{c}$	$\dfrac{\text{deciliters}}{\text{gram}}$	inherent viscosity	logarithmic viscosity number
$[\eta] = \lim\limits_{c \to 0} \eta_{red} = \lim\limits_{c \to 0} \eta_{inh}$	$\dfrac{\text{deciliters}}{\text{gram}}$	intrinsic viscosity	limiting viscosity number

$$\log_{10} \eta_{rel} = \left(\frac{75 k_F^2}{1 + 1.5 k_F c} + k_F \right) c$$

$$c = \text{gram}/\text{dL}$$

$$k_F = \text{Fikentscher constant} \approx \frac{\text{dL}}{\text{gram}}$$

$$K_{value} = 1000 \, k_F$$

Now that the intrinsic viscosity has been established, how is it related to molecular weight? Studies of the intrinsic viscosity of essentially monodisperse polymer fractions whose molecular weights have been established by one of the absolute methods indicate a rather simple relation (Figure 6.4):

$$[\eta]_x = K(M_x)^a (0.5 < a < 1.0) \tag{6-13}$$

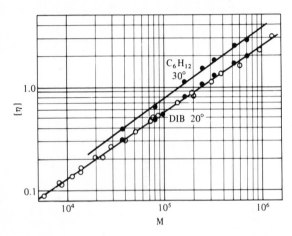

Figure 6.4. Intrinsic viscosity—molecular weight relations for polyisobutylene in cyclohexane at 30C and diisobutylene at 20C. (Reprinted from Paul J. Flory: Principles of Polymer Chemistry. Copyright 1953 by Cornell University. Used by permission of Cornell University Press.)

where the subscript x refers to a monodisperse sample of a particular molecular weight. What about an unfractionated, polydisperse sample? Experimentally, the measured intrinsic viscosity of a mixture of monodisperse fractions is a weight average, i.e.,

$$[\eta] = \frac{\Sigma [\eta]_x w_x}{\Sigma w_x} = \sum \left(\frac{w_x}{W} \right) [\eta]_x. \tag{6-14}$$

A *viscosity-average* molecular weight, $\overline{M}_v$, is defined in terms of this measured intrinsic viscosity:

$$\overline{M}_v = \left(\frac{[\eta]}{K} \right)^{1/a} = \left\{ \frac{\Sigma M_x^a w_x}{W} \right\}^{1/a} = \left\{ \frac{\Sigma M_x^a n_x M_x}{\Sigma n_x M_x} \right\}^{1/a} = \left\{ \frac{\Sigma n_x M_x^{(1+a)}}{\Sigma n_x M_x} \right\}^{1/a}. \tag{6-15}$$

With $0.5 < a < 1.0$, $\overline{M}_n < \overline{M}_v < \overline{M}_w$, but $\overline{M}_v$ is closer to $\overline{M}_w$ than $\overline{M}_n$. If the molecular weight distribution of a series of samples does not differ too much (i.e., if the ratios of the various averages remains the same), approximate equations of the type $[\eta] = K'(\overline{M}_w)^{a'}$ may be applicable.

Example 2. The following data were obtained for a sample of polymethyl methacrylate in acetone at 30C:

η_{rel}	c, g/100 ml
1.170	0.275
1.215	0.344
1.629	0.896
1.892	1.199

For polymethyl methacrylate in acetone at 30C, $[\eta] = 5.83 \times 10^{-5}(\overline{M}_v)^{0.72}$. Determine $[\eta]$ and $\overline{M}_v$ for the sample and k', the constant in the Huggins equation.

Solution. For the data above, calculations give

η_{sp}	$\eta_{red}(dl/g)$	$\ln \eta_{rel}$	$\eta_{inh}(dl/g)$
0.170	0.618	0.157	0.571
0.215	0.625	0.195	0.567
0.629	0.702	0.488	0.545
0.892	0.744	0.653	0.552

In Figure 6.3, η_{red} and η_{inh} are plotted against concentration and extrapolated to a common intercept at zero concentration, the intrinsic viscosity, $[\eta] = 0.577 \, dl/g$:

$$\overline{M}_v = \left(\frac{[\eta]}{5.83 \times 10^{-5}}\right)^{1/0.72} = \left(\frac{0.577}{5.83 \times 10^{-5}}\right)^{1.39} = 355,000.$$

By using the third point in the Huggins Eq. 6-12a, $0.702 \, dl/g = 0.577 \, dl/g + k'(0.577 \, dl/g)^2(0.896 \, g/dl)$. Solving for k' gives $k' = 0.42$ (dimensionless).

Example 3. Assuming the fractions in example 1 are polymethyl methacrylate, calculate $\overline{M}_v$ for mixtures (1) and (2) in acetone at 30C and compare with $\overline{M}_n$ and $\overline{M}_w$.

Solution. From (6-15), for mixture (1)

$$\overline{M}_v = \left\{\sum \left(\frac{w_x}{W}\right) M_x^a\right\}^{1/a}$$

$$= \left\{ \left(\frac{1}{3}\right) (1 \times 10^4)^{0.72} + \left(\frac{2}{3}\right) (4 \times 10^4)^{0.72} \right\}^{1/0.72}$$

$$= 29,000 \qquad\qquad \overline{M}_n = 20,000 \quad \overline{M}_w = 30,000$$

for mixture (2)

$$\overline{M}_v = \left\{ \left(\frac{2}{3}\right) (1 \times 10^4)^{0.72} + \left(\frac{1}{3}\right) (4 \times 10^4)^{0.72} \right\}^{1/0.72}$$

$$= 18,600 \qquad\qquad \overline{M}_n = 13,300 \quad \overline{M}_w = 20,000$$

Note that a is obtained from example 2.

Viscosities for molecular-weight determination usually are measured in glass capillary viscometers in which the solution flows through a capillary under its own head. (Since polymer solutions are non-Newtonian, intrinsic viscosity must be defined, strictly speaking, in terms of the zero-shear or lower-Newtonian viscosity (chapter 15). This is rarely a problem, because the low shear rates in the usual glassware viscometers give just that. Occasionally, however, extrapolation to zero-shear conditions is required.) Two common types, the Ostwald and Ubbelohde are sketched in Figure 6.5. Flow times t are

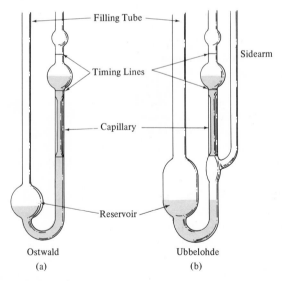

Figure 6.5. Dilute-solution viscometers. (a) Ostwald, (b) Ubbelohde.

related to the viscosity of the solution by an equation of the form

$$\frac{\eta}{\rho} = \iota = at + \frac{b}{t} \tag{6-17}$$

where a and b are instrument constants, ρ the solution density, and ι the kinematic viscosity. The last term, the kinetic energy correction, is generally negligible for flow times of over a minute, and, since the densities of the dilute polymer solutions differ little from that of the solvent,

$$\frac{\eta}{\eta_s} \simeq \frac{t}{t_s}. \tag{6-18}$$

The Ubbelohde viscometer has the distinct advantage that the driving fluid head is independent of the amount of solution in it; hence, dilution of the solutions can be carried out right in the instrument.

The equipment necessary for intrinsic viscosity determination is inexpensive, and the measurements are straightforward and rapid. Constants K and a in (6-13) and (6-15) are extensively tabulated for a wide variety polymer-solvent systems and temperatures (2).

6.4 MOLECULAR-WEIGHT DISTRIBUTIONS

Until fairly recently, the entire molecular-weight distribution of a polymer sample could only be established by laboriously fractionating the sample according to molecular weight and determining the molecular weights of the individual fractions by using one of the techniques just discussed. The fractionation procedures are based on either temperature or solvent power. If a polymer is dissolved in a marginal solvent by heating, the highest molecular weight component will precipitate first as the solution is cooled. Similarly, if a liquid which is not a solvent for the polymer (but is miscible with the solvent) is added slowly to a polymer solution, the high molecular weight material will precipitate first. These fractionation techniques require very dilute solutions (and therefore large volumes of solvent for a small polymer sample), take a long time to reach equilibrium, and present a difficult task in separating the precipitated polymer from the remaining solution. In general, they are avoided whenever possible and have never been commercially feasible for producing monodisperse polymer.

In the past several years, *gel permeation chromatography* (GPC) has come into its own as a means of rapidly determining the entire molecular-weight distribution of a polymer sample. If the distribution is known, any of the averages can be calculated, of course.

Gel permeation chromatography makes use of a column packed with gel (crosslinked polymer which is swelled by the solvent used). The column is equilibrated with the solvent, and then a small amount of a solution of the polymer under investigation is introduced at the top. A constant flow of pure solvent is then initiated, washing the polymer sample downward. The smaller the molecules in the sample are, the more easily they can become entangled in the swollen structure of the gel substrate, and the more difficult will be their progress toward the bottom of the column. The largest molecules, too big to become entrapped in the gel pores, wash through quickly. Thus, a separation is obtained; the largest molecules are washed through the column first, followed by successively smaller ones.

A differential refractometer is placed at the bottom of the column. This gizmo continuously measures the difference in refractive index between the eluant solution and a pure solvent reference. This difference is a very sensitive measure of the solution concentration, so a plot of eluant concentration versus eluant volume is obtained. To obtain the distribution from this, the column must be calibrated by using fractions of known molecular weight to relate molecular weight to eluant volume. Thus, concentration versus molecular weight information is obtained, and it is readily converted to the molecular weight distribution.

Although simple in principle, there are many unanswered questions about GPC, such as the effects of axial and radial dispersion in the column. The technique is currently the subject of intense research interest.

Example 4. Strictly speaking, the number- and weight-average molecular weights should be defined as follows:

$$\overline{M}_n = \frac{\Sigma n_x M_x \Delta M}{\Sigma n_x \Delta M} \quad \text{and} \quad \overline{M}_w = \frac{\Sigma n_x M_x^2 \Delta M}{\Sigma n_x M_x \Delta M}$$

where $\Delta M = M_{x+1} - M_x =$ molecular weight of a repeating unit. Since, for polymers, ΔM is a constant (neglecting end groups), it may be factored out and cancelled, giving (6-2) and (6-3). Even though a molecular weight distribution curve, n_x or (n_x/N) versus M_x or w_x or (w_x/W) versus M_x must be a step function with treads of width ΔM (molecular weights of chains must be integral multiples of ΔM), they are usually approximated by a continuous function for $\overline{M}_w$ and $\overline{M}_n \gg \Delta M$.

Gel permeation chromatography gives the continuous function, $c(M)$ versus M, where c is the *concentration* (weight/volume) of polymer present in the eluant. Given such a function, develop the equations which will allow calculation of $\overline{M}_n$ and $\overline{M}_w$ from the experimental $c(M)$ versus M curve.

$$\text{Remember} \int_a^b f(\xi)\,d\xi = \lim_{\Delta\xi \to 0} \sum_{i=a}^b f(\xi_i)\,\Delta\xi$$

Solution. By applying the limit theorem

$$\overline{M}_w = \frac{\Sigma n_x M_x^2 \Delta M}{\Sigma n_x M_x \Delta M} = \frac{\displaystyle\int_0^\infty n M^2\,dM}{\displaystyle\int_0^\infty n M\,dM} \tag{6-19}$$

and

$$\overline{M}_n = \frac{\displaystyle\int_0^\infty n M\,dM}{\displaystyle\int_0^\infty n\,dM} \tag{6-20}$$

now $n = w/M$ and $c(M) = w/v$, where v is an arbitrarily small volume increment of eluant. Therefore, $n = c(M)v/M$.

$$\overline{M}_w = \frac{\displaystyle\int_0^\infty \frac{c(M)v}{M} M^2\,dM}{\displaystyle\int_0^\infty \frac{c(M)v}{M} M\,dM} = \frac{\displaystyle\int_0^\infty c(M) M\,dM}{\displaystyle\int_0^\infty c(M)\,dM} \tag{6-21}$$

$$\overline{M}_n = \frac{\displaystyle\int_0^\infty \frac{c(M)v}{M}\,M\,dM}{\displaystyle\int_0^\infty \frac{c(M)v}{M}\,dM} = \frac{\displaystyle\int_0^\infty c(M)\,dM}{\displaystyle\int_0^\infty \frac{c(M)}{M}\,dM} \qquad (6\text{-}22)$$

Here, the notation $c(M)$ denotes that c is a function of M, *not* the product of c and M. The integrals above can be evaluated by graphical or numerical techniques.

REFERENCES

1. Billmeyer, F. *Textbook of Polymer Science*, ch. 3. John Wiley & Sons, Inc., New York, 1962.

2. Kurada, M., et al. Viscosity-Molecular Weight Relations, ch. IV-1 in *Polymer Handbook* (J. Brandrup and E. H. Immergut, eds.). John Wiley & Sons, Inc., New York, 1966.

7
Polymer Solubility and Solutions

7.1 INTRODUCTION

The thermodynamics and statistics of polymer solutions is an interesting and important branch of physical chemistry and the subject of many good books or large sections of books in itself. It is far beyond the scope of this chapter to attempt to cover the subject in detail. Instead, we will consider topics of engineering interest and try to indicate, at least qualitatively, their fundamental bases.

Three factors are of general interest to engineers:

1. What solvents will attack what polymers?
2. How does the polymer-solvent interaction influence the solution properties?
3. To what applications do the interesting properties of polymer solutions lead?

7.2 GENERAL RULES FOR POLYMER SOLUBILITY

Let's begin by listing some general experimental observations on the dissolution of polymers:

1. Like dissolves like; i.e., polar solvents will tend to dissolve polar polymers, and nonpolar solvents dissolve nonpolar polymers. Chemical similarity of polymer and solvent is a fair indication of solubility—e.g., polyvinyl alcohol and water,

$$\left[\begin{array}{c} \overset{H}{\underset{|}{C}} - \overset{H}{\underset{|}{C}} \\ \overset{|}{H} \quad \overset{|}{OH} \end{array}\right]_x$$ and $H\diagup\overset{O}{}\diagdown H$, or polystyrene and toluene,

$$\left[\begin{array}{c} \overset{H}{\underset{|}{C}} - \overset{H}{\underset{|}{C}} \\ \overset{|}{H} \quad \bigcirc \end{array}\right]_x$$ and $\overset{CH_3}{\bigcirc}$

2. In a given solvent at a particular temperature, the solubility of a polymer will decrease with increasing molecular weight.

3. (a) Crosslinking eliminates solubility. (b) Crystallinity, in general, acts like crosslinking, but it is possible in some cases to find solvents strong enough to overcome the crystalline bonding forces and dissolve the polymer. Heating the polymer above its crystalline melting point allows its solubility in appropriate solvents.

4. The *rate* of polymer solubility (a) increases with short branches, which loosen up the main-chain structure, allowing the solvent molecules to penetrate more easily, and (b) decreases with longer branches, because the entanglement of these branches makes it harder for individual molecules to separate.

It is important to note here that items 1, 2 and 3 are *equilibrium* phenomena and are therefore describable thermodynamically (at least in principle), while 4 is a *rate* phenomenon and is governed by the rates of diffusion of polymer and solvent.

Example 1. The polymers of ω-amino acids are termed nylon n, where n is the number of consecutive carbon atoms in the chain. Their general formula is:

$$\left[\begin{array}{c} \overset{H}{\underset{|}{N}} - \overset{O}{\underset{||}{C}} - \left(CH_2\right)_{n-1} \end{array}\right]_x.$$

The polymers are crystalline and will not dissolve in either water or hexane $\left(H_3C-\left(CH_2\right)_4-CH_3\right)$ at room temperature. They will, however, reach an equilibrium level of absorption when immersed in each liquid. Describe how and why water and hexane absorption will vary with n.

Solution. Water is a highly polar liquid; hexane is nonpolar. The polarity of the nylons depends on the relative proportion

of polar nylon linkages $\left(\begin{array}{c} H \quad O \\ | \quad \| \\ N-C \end{array}\right)$ in the chains. As n increases, the polarity of the chains decreases (they become more hydrocarbonlike), and so hexane absorption increases with n and water absorption decreases.

7.3 THE THERMODYNAMIC BASIS OF POLYMER SOLUBILITY

To dissolve or not to dissolve; that is the question (with apologies to W. Shakespeare). The answer is determined by the sign of the Gibbs free energy. Consider the process of mixing pure polymer and pure solvent (state 1) at constant pressure and temperature to form a solution (state 2).

$$\Delta F = \Delta H - T \Delta S \qquad (7\text{-}1)$$

where ΔF = the change in Gibbs free energy in the process,

ΔH = the change in enthalpy in the process,

T = the absolute temperature at which the process is carried out,

and ΔS = the change in entropy in the process.

Only if ΔF is negative will the solution process be thermodynamically feasible. The absolute temperature must be positive, and the change in entropy for a solution process is positive because, in a solution, the molecules are in a more random state than in the solid. The positive product is preceded by a negative sign. Thus the third $(-T\Delta S)$ term in (7-1) favors solubility. The change in enthalpy may be either positive or negative. A positive ΔH means the solvent and polymer prefer their own company (i.e., the pure materials are in a lower energy state), while a negative ΔH indicates that the solution is the lower energy state. If the latter obtains, solution is assured. Negative ΔH's occur where specific interactions (e.g., hydrogen bonds) are formed between the solvent and polymer molecules. Thus, if ΔH is positive, $\Delta H < T\Delta S$ if the polymer is to be soluble.

The entropy change in forming a polymer solution is quite small—orders of magnitude smaller than that which occurs when equivalent masses or volumes of two low molecular-weight liquids are mixed. The reasons for this are illustrated qualitatively on a two-dimensional lattice model in Figure 7.1

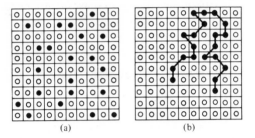

Figure 7.1.   Lattice model of solubility.   Open circles = solvent; filled circles = solute. (a) low molecular weight solute, (b) polymeric solute.

(such models formed the basis for early quantitative treatments of polymer solubility).   In the low molecular-weight mixture, the solute molecules may be distributed randomly throughout the lattice with the only restriction being that two lattice sites cannot be occupied simultaneously. This gives rise to a large number of configurational possibilities—i.e., a high entropy. In the polymer solution, however, each chain segment is confined to the lattice site adjacent to the next chain segment, greatly reducing the configurational possibilities.   Note also that, for a given number of chain segments (equivalent masses or volumes of polymer), the more chains they are split up into (i.e., the lower the molecular weight), the higher will be their entropy upon solution.   This directly explains observation 2, the decrease in solubility with molecular weight. However, in general, for high molecular weight polymers, because the $T\Delta S$ term is so small, if ΔH is positive, it must be even smaller— pretty near zero, if the polymer is to be soluble.

7.4 PREDICTION OF SOLUBILITY

How can ΔH be estimated?   Well, for *regular* solutions (those in which solute and solvent *do not* form specific interactions), the change in *internal energy* upon solution is given by

$$\Delta H \simeq \Delta E = \phi_1 \phi_2 (\delta_1 - \delta_2)^2 \left(\frac{cal}{cc \; solution} \right) \qquad (7\text{-}2)$$

where ΔE = the change in internal energy in the solution process,

ϕ = volume fractions,

and δ = *solubility parameters*.

The subscripts 1 and 2 usually refer to solute (polymer) and solvent, respectively. The *solubility parameter* is defined as follows:

$$\delta = (CED)^{1/2} = \left(\frac{\Delta E_v}{v} \right)^{1/2} \qquad (7\text{-}3)$$

where (CED) = *cohesive energy density*, a measure of the strength of the intermolecular forces holding the molecules together in the liquid state,

ΔE_v = molar change in internal energy on vaporization,

and v = molar volume of liquid.

Now, for a process which occurs at constant volume and constant pressure, the changes in internal energy and enthalpy are equal. This is a good approximation for the dissolution of polymers under most conditions, so (7-2) provides a means of estimating enthalpies of solution if the solubility parameters of the polymer and solvent are known.

Note that, regardless of the magnitudes of δ_1 and δ_2 (they are always positive), the predicted ΔH is always positive; i.e., (7-2) applies only in the absence of specific interactions which lead to negative ΔH's. Inspection of (7-2) also reveals that ΔH is minimized, and the tendency toward solubility maximized by matching the solubility parameters as closely as possible.

Measuring the solubility parameter of a low molecular-weight solvent is no problem. Polymers, on the other hand, degrade long before reaching their vaporization temperatures, making it impossible to evaluate ΔE_v directly. Fortunately, there is a way around this impasse. The greatest tendency of a polymer to dissolve occurs when its solubility parameter matches that of the solvent. If the polymer is crosslinked lightly, it cannot dissolve but only swell. The maximum swelling will be observed when the polymer and solvent solubility parameters are matched. So polymer solubility parameters are determined by soaking lightly crosslinked samples in a series of solvents of known solubility parameters. The value of the solvent at which maximum swelling is observed is taken as the solubility parameter of the polymer (Figure 7.2). Solubility

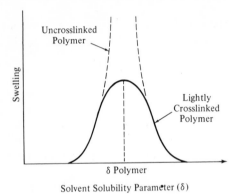

Figure 7.2.   Determination of polymer solubility parameter by swelling lightly crosslinked samples in a series of solvents.

parameters of solvent mixtures can be readily calculated

$$\delta_{mixture} = \frac{x_1 v_1 \delta_1 + x_2 v_2 \delta_2}{x_1 v_1 + x_2 v_2}$$ (7-4)

where x = mole fraction.

As a further aid in predicting the susceptibility of polymers to various solvents, techniques have been described which supplement solubility parameters with quantitative information on hydrogen bonding and dipole moments (1,2).

7.5 PROPERTIES OF DILUTE SOLUTIONS

Well, now the polymer is in solution. Let's assume it's a fairly dilute solution so we don't have to worry about too many entanglements between the molecules. In a good solvent (one whose solubility parameter closely matches that of the polymer), the secondary forces between polymer segments and solvent molecules are strong, and the polymer molecules will assume a spread out conformation in solution. In a poor solvent, the attractive forces between the segments of the polymer chain will be greater than those between the chain segments and the solvent; i.e., the chain segments prefer their own company, and the chain will ball up tightly (Figure 7.3).

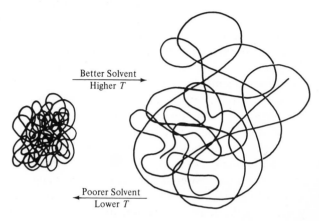

Better Solvent
Higher T

Poorer Solvent
Lower T

Figure 7.3.   The effects of solvent power and temperature on a polymer molecule in solution.

Imagine a polymer in a good solvent—e.g., polystyrene (δ = 8.56) in carbon tetrachloride (δ = 8.6). A nonsolvent is now added—e.g., methanol (δ = 14.5). (Despite the large difference in solubility parameters, carbon tet and methanol are mutually soluble in all proportions because of the large ΔS of solution for low molecular weight compounds.) Ultimately, a point is reached where the mixed solvent becomes too poor to sustain solution, and the polymer precipitates out as the attractive force between polymer segments becomes much greater than that between polymer and solvent. At some point, the polymer teeters on the brink of solubility; i.e., $\Delta F = 0$, and $\Delta H = T\Delta S$. This point obviously depends on the temperature, polymer molecular weight (mainly through its influence on ΔS) and the polymer-solvent system (mainly through its influence on ΔH). Adjusting either the temperature or the polymer-solvent system allows fractionation of the polymer according to molecular weight, as successively smaller molecules precipitate upon lowering the temperature or going to poorer solvent. In the limit of infinite molecular weight (the minimum possible ΔS), the situation where $\Delta H = T\Delta S$ is known as the 'θ condition.' Under these conditions, the polymer-solvent and polymer-polymer interactions are equal, and the solution behaves in a so-called ideal fashion, with the second virial coefficient equal to 0, etc. For a given polymer, the θ condition can be reached at a fixed temperature by adjusting the solvent to give a θ solvent or, with a particular solvent, by adjusting the temperature to reach the θ or Flory temperature. Any actual polymer will still be soluble under θ conditions, of course, because of its lower-than-infinite molecular weight and consequently larger ΔS.

Getting back to our example, we might ask what happens to the viscosity of the solution upon going from a good solvent to a poor solvent (to make a fair comparison, imagine that, as nonsolvent is added, an equivalent amount of good solvent is removed, maintaining constant the polymer concentration). This question can be answered qualitatively by imagining the polymer molecules in solution to be rigid spheres (they aren't) and by applying the Einstein relation, (6-7) (further assuming that the viscosities of the low molecular weight liquids are comparable—i.e., that η_s doesn't change). When they go from good to poor solvent, the polymer molecules ball up, giving a smaller effective ϕ and *lowering* the solution viscosity. Thus, solution viscosity can be controlled by adjusting solvent power. This fact is of great importance to the surface-coatings industry. For example, in formulating a lacquer (a solution of polymer in solvent, plus some pigment, which dries only by solvent evaporation), the viscosity for optimum spraying or brushing characteristics might be obtained with a mixed solvent, with the poorer component the more volatile. After application, the poorer component evaporates first, leaving a high viscosity and, therefore, sag- and run-resistant film on the substrate.

Example 2. Indicate how solvent power (good solvent versus poor solvent) will influence the following:

a. the design of a stirred solution polymerization reactor;

b. the intrinsic viscosity of a polymer sample at a particular T; and

c. the molecular weight of a polymer sample as determined by membrane osmometry.

Solution

a. As long as the polymer remains in solution, a poor solvent, with its lower viscosity, will permit better agitation or require a less powerful stirrer motor.

b. The measured viscosities of solutions will be less in a poor solvent than in a good one, leading to lower intrinsic viscosities.

c. If the membrane were ideal, solvent power would make no difference, as osmometry measures only the number (moles) of solute per unit volume, regardless of geometry. Some of the smallest particles can sneak through any real membrane, however. The poor solvent, causing the molecules to ball up tightly, will allow passage of more molecules through the membrane. This will lower the observed osmotic pressure, causing the calculated $\overline{M}_n$ to err on the high side.

Consider now the effects of temperature on the viscosity of a solution of polymer in a relatively poor solvent. As is the case with all simple liquids, the solvent viscosity, η_s, decreases with increasing temperature. Increasing temperature, however, imparts more thermal energy to the segments of the polymer molecules, causing the molecules to spread out and assume a larger effective ϕ in solution. Thus, the effects of temperature on η_s and ϕ tend to compensate, giving a solution which has a much smaller change in viscosity with temperature than that of the solvent alone. In fact, the additives which are used to produce the so-called multiviscosity (10W-30) motor oils are nothing more than polymers, with compositions adjusted so that the base oil is a relatively poor solvent at the lowest operating temperatures. As the engine heats up, the polymer molecules uncoil, providing a much greater resistance to thinning than is possible from oil alone.

Viscosities (at low shear rates) of dilute solutions of polymers with known molecular weights may be calculated by using (6-12) and (6-13), reversing the procedure for obtaining molecular weights from viscosity measurements.

7.6 POLYMER-POLYMER-COMMON SOLVENT SYSTEMS

We have discussed the small increase in entropy which arises when a high molecular weight polymer is dissolved in a solvent.

Applying the same reasoning to the dissolution of one high molecular weight polymer in another, the ΔS of solution will be even smaller. For this reason, the true solubility of one polymer in another is extremely rare. Even when a common solvent is added (one which is infinitely soluble with each polymer alone), the two polymers cannot coexist in a homogeneous phase beyond a few per cent concentration. A schematic phase diagram for such a system is shown in Figure 7.4. Be-

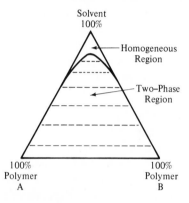

Figure 7.4. Typical ternary phase diagram for a polymer-polymer-common solvent system. The dotted tie lines connect ·compositions of phases in equilibrium.

yond a few per cent polymer (the exact value depends on the chemical nature of the polymers and solvent and the molecular weights of the polymers), two phases in equilibrium are formed with each phase containing a nearly pure polymer. These results extend to more than two polymers. In general, each polymer will coexist in a separate phase.

7.7 CONCENTRATED SOLUTIONS— PLASTICIZERS

Up to this point, we have considered relatively dilute polymer solutions. Now let's look at the other end of the spectrum where the polymer is the major constituent of the solution. A pure, amorphous polymer consists of a tangled mass of polymer chains. The ease with which this mass can deform depends on the ability of the polymer chains to untangle and slip past one another. One way of increasing this ability is to raise the temperature, as we have seen. Another is to add a low molecular-weight liquid, an *external plasticizer*, to the polymer. By forming secondary bonds to the polymer molecules and spreading them apart, the plasticizer reduces the polymer-polymer chain secondary bonding, and it provides more room for the polymer molecules to move around, providing a softer, more easily deformable mass.

Example 3. If you have ever heard baseball announcers comment that "On these humid nights, the ball just isn't hit as hard" or "A pitcher throws a heavier ball on these humid nights," you may have wondered if there was a scientific basis for these statements. They cannot be justified in terms of the properties of the humid air (fluid dynamics). Can you justify them in terms of the baseball properties? A baseball is made largely of tightly wound wool yarn. Wool, a natural polymer, is similar, physically and chemically, to nylons (see example 1).

Solution. The polar linkages will hydrogen bond to water. The higher the humidity is, the greater the equilibrium moisture content of the wool will be, accounting for the heavier ball. The absorbed water will also act as an external plasticizer for the wool, softening it and decreasing its resiliency.

A poor (from the solvent standpoint) plasticizer should be more effective than a good one, giving a lower viscosity at a given level (smaller ϕ, fewer entanglements), but, since it is less strongly bound to the polymer, it will have a greater tendency to exude out over a period of time, leaving the polymer mass stiffer. This was a familiar problem with early shower curtains, etc. Thus, a balance must be struck between plasticizer efficiency and permanence. Also, from the standpoint of permanence, it is necessary that the plasticizer have low volatility. They therefore generally have higher molecular weights than solvents, but are still well below the high polymer range in this respect.

A polymer may be *internally plasticized* by copolymerization with a low-T_g monomer (see chapter 8). There are obviously never any permanence problems when this is done. The composition of the copolymer can be adjusted to give the desired properties at a particular temperature. Above this temperature, the copolymer will be softer than intended; below, it will be harder. With external plasticizers, the same sort of temperature compensation as in the multivis motor oils is obtained, giving materials which maintain the desired flexibility over a broader temperature range.

The most common externally plasticized polymer is poly vinyl chloride (PVC)

$$\left[\begin{array}{c} H \\ | \\ C \\ | \\ H \end{array} - \begin{array}{c} H \\ | \\ C \\ | \\ Cl \end{array}\right]_x.$$

A typical plasticizer for PVC is dioctyl phthalate (DOP), the ester of phthalic acid and octyl alcohol

Phosphates (e.g., tricresyl phosphate, TCP), as well as chlorinated waxes, are used as plasticizers which impart fire resistance. So-called polymeric plasticizers, polymers with molecular weights on the order of 1000, provide low volatility and good permanence. It might be noted that a low molecular weight fraction in a pure polymer behaves as a plasticizer, often in an undesirable fashion.

Epoxidized plasticizers are becoming increasingly important. In addition to plasticizing PVC, these compounds also perform the important function of stabilizing the resin. When PVC degrades, HCl is given off, and the HCl catalyzes further degradation. It also attacks metallic molding machines, molds, extruders, etc. The epoxy plasticizers are made from unsaturated oils

oxirane ring

The *oxirane* rings soak up HCl and minimize further degradation.

$$\underset{\underset{O}{\diagdown\diagup}}{\overset{H\quad H}{\underset{|\quad\;|}{\sim\!\!\sim\!C-C\sim\!\!\sim}}} + HCl \rightarrow \underset{\overset{|\quad\;|}{OH\;Cl}}{\overset{H\quad H}{\underset{|\quad\;|}{\sim\!\!\sim\!C-C\sim\!\!\sim}}}$$

Unplasticized (or nearly so) PVC is a rigid material used for pipe and fittings among other things. The properties of plasticized PVC vary considerably depending on the plasticizer level. It is familiar as a gasketing material, as a leather-like upholstery material, a wire and cable covering, shower curtains, etc.

Plastisols are an interesting and useful technological application of plasticized PVC. A typical formulation might consist of 100 parts DOP phr (per hundred parts resin) plus some stabilizers, pigments, etc. Although the PVC is thermodynamically soluble in the DOP at room temperature, the rate of dissolution is extremely slow. Thus, initially, the plastisol is a milky suspension of finely divided resin particles in the plasticizer. As a suspension rather than a solution, the viscosity is of the order of magnitude of that of the plasticizer itself, and it can be applied to substrates or molds by brushing, dipping, rolling, etc. When heated to about 350F, the increased thermal agitation of the polymer molecules speeds up the solution process greatly. If there is no filler or pigment present, the solution process can be observed as the plastisol becomes transparent. When cooled back to room temperature, the viscosity of the solution is so high that, for all practical purposes, it may be considered a flexible solid. Examples of plastisols include doll 'skin' and the covering on wire dish racks. The viscosity of the initial suspension may be lowered by incorporating a volatile organic diluent for the plasticizer, giving an *organosol*, or by whipping in water to form a *hydrosol*. In both cases, the diluent vaporizes upon heating.

REFERENCES

1. Beerbower, A., L. A. Kaye and D. A. Pattison. Picking the right elastomer to fit your fluids. *Chem. Eng.*, **74,** no. 26, p. 118, 1967.

2. Hansen, C. M. The three-dimensional solubility parameter. *J. Paint Technol*, **39,** no. 505, p. 104, 1967.

8
Transitions in Polymers

8.1 THE GLASS TRANSITION

It has long been known that amorphous polymers can exhibit two distinctly different types of mechanical behavior. Some, like polystyrene and polymethyl methacrylate (Lucite and Plexiglass), are hard, rigid, *glassy* plastics at room temperature, while others (e.g., polybutadiene, polyethyl acrylate and polyisoprene) are soft, flexible rubbery materials. If, however, polystyrene and polymethyl methacrylate are heated to about 125C, they exhibit typical rubbery properties, and, when a rubber ball is cooled in liquid nitrogen, it becomes rigid and glassy, and it shatters when an attempt is made to bounce it. Therefore, there is some temperature, or narrow range of temperatures, below which an amorphous polymer is in a glassy state and above which it is rubbery. This temperature is known as the *glass transition temperature*, T_g. The glass transition temperature is a property of the polymer, and whether the polymer has glassy or rubbery properties depends on whether its application temperature is above or below its glass transition temperature.

8.2 MOLECULAR MOTIONS IN AN AMORPHOUS POLYMER

In order to understand the molecular basis for the glass transition, the various molecular motions occurring in an amorphous polymer mass may be broken into four categories:

1. translational motion of entire molecules, which permits flow;

2. cooperative wriggling and jumping of segments of molecules approximately 40–50 carbon atoms in length, permitting flexing and uncoiling;

3. motions of a few atoms along the main chain (five or six, or so) or of side groups on the main chains; and

4. vibrations of atoms about equilibrium positions, as occurs in crystal lattices, except that the atomic centers are not in a regular arrangement in an amorphous polymer.

The motions 1–4 above are arranged in order of decreasing activation energy; i.e., smaller amounts of thermal energy (kT) are required to produce them. The glass transition temperature is thought to be that temperature at which motions 1 and 2 are pretty much frozen out, and there is only sufficient energy available for motions of types 3 and 4. Of course, not all molecules possess the same energies at a given temperature. The molecular energies follow a Boltzmann distribution, and, even below T_g, there will be occasional type 2 motions which can manifest themselves over extremely long periods of time.

8.3 DETERMINATION OF T_g

How is the glass transition studied? A common method is to observe the variation of some thermodynamic property with T, for example, the specific volume, as shown in Figure 8.1. Note

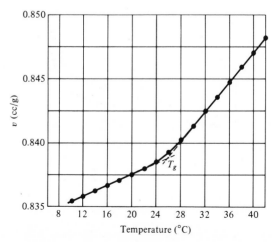

Figure 8.1. Specific volume vs. temperature for polyvinyl acetate (2).

that the slope of the v versus T plot increases above the glass transition temperature.

The value of T_g determined in this fashion will vary slightly with the rate of cooling or heating. This reflects the fact that long, entangled polymer chains cannot respond instantaneously to changes in temperature and illustrates the difficulty in making thermodynamic measurements on polymers. It often takes extremely long to reach equilibrium, if indeed it is ever reached, and it is difficult to be sure if and when it is reached. Strictly speaking, the glass transition temperature should be defined in terms of equilibrium properties, or at least those measured with very low rates of temperature change. Also, there is never observed a sharp break in the property, but this is of little practical importance, because T_g can always be established within a couple of degrees by extrapolation of the linear regions. Other properties such as refractive index may also be used to establish T_g.

In contrast to a change in *slope* at the glass transition, a thermodynamic property such as specific volume exhibits a discontinuity with temperature at the crystalline melting point in polymers as in other materials (Figure 8.2). The glass tran-

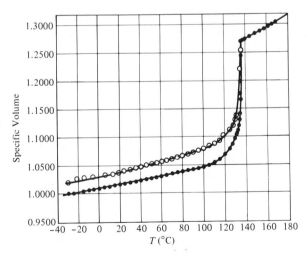

Figure 8.2.   Specific volume-temperature relations for linear polyethylene (Marlex-50). Specimen slowly cooled from melt to room temperature prior to fusion, open circles; specimen crystallized at 130C for forty days then cooled to room temperature prior to fusion, closed circles (*3*).

sition is therefore known as a *second-order* thermodynamic transition (v versus T continuous, dv/dT versus T discontinuous) in contrast to a first-order transition such as the melting point (v versus T discontinuous). There is still considerable argument as to whether T_g is a true second-order transition, but this is of little practical consequence. Since all polymers have at least some amorphous material (they can't be 100 per cent crystalline), they all have a T_g, but not all polymers have a

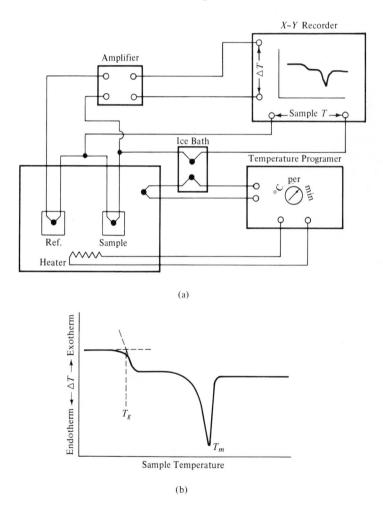

(a)

(b)

Figure 8.3.   Differential thermal analysis. (a) schematic of apparatus, (b) DTA curve.

crystalline melting point—they can't if they don't have any crystallinity.

A new and powerful tool for analyzing polymer transitions is differential thermal analysis (DTA). Small samples of the polymer and an inert reference substance (one which undergoes no transition in the temperature range of interest) are mounted in a block with thermocouples to monitor temperatures (Figure 8.3). The thermodynamic property monitored here is the enthalpy. The block is heated at a constant rate, and the *difference* between the sample and reference temperatures is plotted versus the sample temperature. At T_g, $C_p = \left(\dfrac{\partial H}{\partial T}\right)_p$ increases (the slope of an H versus T plot increases), dropping the sample temperature below its previous level relative to the reference. At T_m, large amounts of heat are required to melt the crystals at constant T, and the ΔT shows a sharp dip. This technique can also be used to follow oxidation and degradation reactions.

8.4 FACTORS INFLUENCING T_g (1)

In general, the glass transition temperature depends on five factors.

1. The *free volume* of the polymer, v_f Free volume is the volume of the polymer mass not actually occupied by the molecules themselves; i.e., $v_f = v - v_s$, where v is the specific volume of the polymer mass and v_s is the volume of the solidly packed molecules. The higher v_f is, the more room the molecules will have in which to move around, and the lower will be T_g. It has been estimated that for all polymers $(v_f/v) = 0.025$ at T_g.

Example 1. Glass-transition temperatures have been observed to increase at pressures of several thousand psi. Why?

Solution. High pressures compress polymers, reducing v. Since v_s doesn't change appreciably, v_f is reduced.

2. The attractive forces between the molecules The more strongly they are bound together, the more thermal energy will be required to produce motion. Since the solubility parameter, δ, is a measure of intermolecular forces, T_g increases with δ.

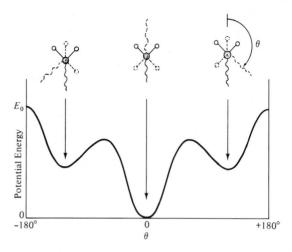

Figure 8.4.   Rotation about a bond in a polymer chain backbone, viewed along the bond. The dotted substituents are on the rear carbon atom.

3. The internal mobility of the chains—i.e., their freedom to rotate about bonds

Figure 8.4 shows potential energy as a function of rotation angle about a bond in a polymer chain. The minimum energy configuration, arbitrarily chosen as $\theta = 0$, is the position where the largest substituents, the rest of the chain, are as far away from each

TABLE 1   (1)

polymer		δ	T_g	E_o
Silicone rubber	$\left[\begin{array}{c} CH_3 \\ \mid \\ -Si-O- \\ \mid \\ CH_3 \end{array}\right]_x$	7.3	$-120C$	$\sim 0\ Kcal/mol$
Polyethylene	$\left[\begin{array}{c} H\quad H \\ \mid\quad \mid \\ -C-C- \\ \mid\quad \mid \\ H\quad H \end{array}\right]_x$	7.9	-85	3.3
Polytetrafluoroethylene	$\left[\begin{array}{c} F\quad F \\ \mid\quad \mid \\ -C-C- \\ \mid\quad \mid \\ F\quad F \end{array}\right]_x$	6.2	>20	4.7

other as possible. As the bond is rotated, the substituent groups are brought into juxtaposition, and energy is required to push them over the hump. The maximum energy is needed to get the two chain substituents past one another, and this energy must be available if complete rotation is to be obtained. Table 1 shows how T_g increases with E_o for a series of polymers with approximately the same δ. Note how the ether oxygen "swivel" in the silicone chain permits very free rotation.

Example 2. Poly α-methylstyrene has a higher T_g than polystyrene. Why?

polystyrene   poly α-methylstyrene

Solution. The methyl group introduces extra steric hindrance to rotation, giving a higher E_o.

4. The stiffness of the chains   Chains which have difficulty coiling and folding will have higher T_g's. This stiffness usually goes hand-in-hand with high E_o, so it is difficult to separate the effects of 3 and 4.

5. The chain length   As do many mechanical properties of polymers, the glass transition temperature varies according to the empirical relation

$$T_g = T_g^\infty - \frac{C}{x} \tag{8-1}$$

where C is a constant for the particular polymer, and T_g^∞ is the asymptotic value of the glass transition temperature at infinite chain length. In most cases, for $x > 500$ (the range of commercial interest for most polymers), $T_g \simeq T_g^\infty$.

Example 3. Measurements such as those described above show that an external plasticizer softens a polymer by reducing its glass-transition temperature. Explain.

Solution. The plasticizer molecules pry apart the polymer chains, in essence increasing the free volume available to the chains (although not truly free, the small plasticizer molecules interfere with chain motions much less than would other chains). Also, by forming secondary bonds with the polymer chains, the plasticizer molecules reduce the bonding forces between the chains themselves.

8.5 T_g's OF COPOLYMERS

The glass transition temperatures for random copolymers vary monotonically with composition between those of the homopolymers. They can be approximated fairly well from a knowledge of the T_g's of the homopolymers with the empirical relation

$$\frac{1}{T_g} = \frac{w_1}{T_{g1}} + \frac{w_2}{T_{g2}} \tag{8-2}$$

where the w's are weight fractions of the monomers in the copolymer. This relation forms the basis for a method of estimating the T_g's of highly crystalline polymers, where the properties of the small amount of amorphous material are masked by the majority of crystalline material present. If a series of random copolymers can be produced in which the randomness prevents crystallization over a certain composition range, (8-2) can be used to extrapolate to $w_1 = 0$ and $w_1 = 1$, giving the T_g's of the homopolymers. This method is open to question, because it assumes that the presence of major amounts of crystallinity does not restrict the molecular response in the amorphous regions.

8.6 THE THERMODYNAMICS OF MELTING

The crystalline melting point, T_m, in polymers is a phase change similar to that observed in low molecular weight organic compounds, metals and ceramics.

The (Gibbs) free energy of melting is given by

$$\Delta F_m = \Delta H_m - T\Delta S_m. \tag{8-3}$$

At the crystalline melting point, T_m, $\Delta F_m = 0$, so

$$T_m = \Delta H_m / \Delta S_m. \tag{8-4}$$

Now ΔH_m is the energy needed to overcome the crystalline bonding forces at constant T and P, and it is essentially independent of chain length for high polymers. For a given mass or volume of polymer, however, the shorter the chains are, the more randomized they become upon melting, giving a higher ΔS_m. (For a more detailed description of this in connection with solutions, see Chapter 7.) Thus, the crystalline melting point decreases with decreasing chain length, and, in a polydisperse polymer, the distribution of chain lengths gives a distribution of melting points or rounding off noted in Figure 8.2.

Equation (8-4) also indicates that chains which are strongly bound in the crystal lattice (i.e., have a high ΔH_m) will have a high T_m, as expected. Also, the stiffer and less mobile chains, those which can randomize less upon melting and therefore have a low ΔS_m, will tend to have higher T_m's.

Example 4. Consider the following classes of linear, aliphatic polymers.

$$\left[O-\overset{\overset{\displaystyle O}{\parallel}}{C}-\underset{\underset{\displaystyle H}{|}}{N}-(CH_2)_n \right]_x \qquad \left[\overset{\overset{\displaystyle O}{\parallel}}{C}-\underset{\underset{\displaystyle H}{|}}{N}-(CH_2)_n \right]_x$$

polyurethanes   polyamides
T_m < T_m

$$\left[\underset{\underset{\displaystyle H}{|}}{N}-\overset{\overset{\displaystyle O}{\parallel}}{C}-\underset{\underset{\displaystyle H}{|}}{N}-(CH_2)_n \right]_x$$

<   polyureas
T_m

For given values of n and x, the crystalline melting points increase from left to right, as indicated. Explain.

Solution. The polyurethane chains contain the —O— swivel; thus, they are the most flexible, have the largest ΔS_m and the lowest T_m. Hydrogen bonding, and thus ΔH_m, is roughly comparable in the polyurethanes and polyamides. The polyureas and polyamides should have chains of comparable flexibility (no swivel), but, with the extra —N— , the polyureas form

$$\overset{\displaystyle H}{\underset{\displaystyle |}{}}$$

stronger hydrogen bonds and therefore have a higher ΔH_m than the polyamides.

The crystalline melting point also increases a bit with the degree of crystallinity of a polymer. For example, low-density polyethylene (about 65 per cent crystalline) has a T_m of about 115C, whereas high density polyethylene (about 95 per cent crystalline) melts at about 135C. This can be explained by treating the amorphous material as an impurity. It is well known that the introduction of an impurity lowers the melting point of common materials. In a similar fashion, greater amounts of noncrystalline impurities lower the crystalline melting point of a polymer.

Since polymer chains are largely immobilized below T_g, if they are cooled rapidly through T_m to below T_g, it is often possible to obtain normally crystalline polymers in a metastable amorphous state which will persist indefinitely. When annealed above T_g (and below T_m), they will crystallize, as the chains gain the mobility necessary to pack into a lattice.

Example 5. Polyethylene terephthalate (Mylar, Dacron) is cooled rapidly from 300C (state 1) to room temperature. The resulting material is rigid and perfectly transparent (state 2). The sample is then heated to 100C and maintained at that temperature, during which time it gradually becomes translucent (state 3). It is then cooled down to room temperature and is again found to be rigid, but it is now translucent rather than transparent (state 4). For this polymer, $T_m = 267C$, and $T_g = 69C$.

Sketch a general specific-volume versus temperature curve for a crystallizable polymer, illustrating T_g and T_m, and show the locations of states 1–4 for the sample above.

Solution. Figure 8.5 illustrates the general v versus T curve for a crystallizable polymer. The dotted upper portion represents the metastable amorphous material obtainable by rapid cooling. The history above is shown on the diagram. The metastable amorphous material (transparent, state 2) obtained by rapid cooling to below T_g crystallizes (to translucent, state 3) upon annealing between T_g and T_m.

The greater mobility of chains crystallized just below T_m accounts for their higher degree of crystallinity observed in Figure 8.2.

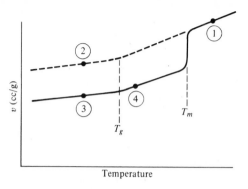

Figure 8.5. Specific volume-temperature relation for crystallizable polymer. Numbers apply to example 5.

8.7 THE INFLUENCE OF COPOLYMERIZATION ON PROPERTIES

The influence of random copolymerization on T_m and T_g is interesting and technologically important. Occasionally, if two monomers are similar enough sterically to fit into the same crystal lattice, random copolymerization will result in co-polymers whose crystalline melting points vary monotonically with composition between those of the pure homopolymers. More common, however, is the case where the homopolymers form different crystal lattices because of steric differences. The random copolymerization of minor amounts of monomer B with A will disrupt the A lattice, lowering the T_m beneath that of homopolymer A, and vice versa. In an intermediate com-position range, the disruption will be so great that no crystal-lites can form, and the copolymers will be completely amor-phous. A phase diagram for such a random copolymer is shown in Figure 8.6.

The physical properties of random copolymers are de-termined by the region of the phase diagram—i.e., by their composition and use temperature. In region 1, the polymer is a homogeneous, amorphous, and (if pure) transparent mate-rial. The distinction between melt and rubbery behavior is not sharp—at higher temperatures, the material flows more easily and becomes less elastic in character. It should be kept

86

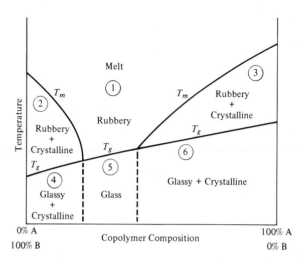

Figure 8.6.   Phase diagram for a random copolymer system.

in mind, though, that the viscosities of polymer melts, even well above T_m or T_g, are far greater than those encountered in nonpolymeric materials.   A typical value might be on the order of a million centipoise.

A copolymer in region 5 is a typical amorphous, glassy polymer, hard, rigid and usually brittle.   Again, if the polymer is pure, it will be perfectly transparent.   Polystyrene and polymethyl methacrylate (Lucite, Plexiglas) are familiar examples of homopolymers with these properties.

Copolymers in regions 2 and 3 consist of rigid crystallites dispersed in a relatively soft, rubbery amorphous matrix.   Since the refractive indices of the crystalline and amorphous phases differ, materials in these regions will be translucent to opaque, depending on the degree of crystallinity and the thickness of the sample.   Since the crystallites restrict chain mobility, the materials are not elastic, but the rubbery matrix confers flexibility and toughness.   The stiffness depends largely on the degree of crystallinity—the more rigid crystalline phase present, the stiffer the polymer.   Polyethylene (squeeze bottles, bleach bottles, etc.) is a good example of a homopolymer with these properties—i.e., one which is *between* its T_g and T_m at room temperature.

Copolymers in regions 4 and 6 consist of crystallites in an

amorphous, glassy matrix. Since both phases are rigid, the materials are hard, stiff, and rigid. Again, the two phases impart opacity. Nylons 6-6 and 6 are examples of homopolymers in a region below both T_g and T_m at room temperature.

8.8 GENERAL OBSERVATIONS ABOUT T_g AND T_m

Some other useful observations regarding T_m and T_g are that, for polymers with a symmetrical repeating unit (e.g.,

polyethylene $\left[\begin{matrix} \text{H} & \text{H} \\ | & | \\ \text{C} - \text{C} \\ | & | \\ \text{H} & \text{H} \end{matrix}\right]_x$ and polyvinylidene chloride (Saran)

$\left[\begin{matrix} \text{H} & \text{Cl} \\ | & | \\ \text{C} - \text{C} \\ | & | \\ \text{H} & \text{Cl} \end{matrix}\right]_x$, $T_g/T_m \cong 1/2$ and for unsymmetrical repeating

units $\Bigg($e.g., polypropylene $\left[\begin{matrix} \text{H} & \text{CH}_3 \\ | & | \\ \text{C} - \text{C} \\ | & | \\ \text{H} & \text{H} \end{matrix}\right]_x$ and polychlorotri-

fluoroethylene (Kel-F) $\left[\begin{matrix} \text{F} & \text{F} \\ | & | \\ \text{C} - \text{C} \\ | & | \\ \text{F} & \text{Cl} \end{matrix}\right]_x\Bigg)$, $T_g/T_m \cong 2/3$.   In all

cases $T_g < T_m$.

8.9 EFFECTS OF CROSSLINKING

To this point, the discussion has centered on noncrosslinked polymers. Light crosslinking, as in rubber bands, won't alter things appreciably. Higher degrees of crosslinking, however, if formed in the amorphous melt state as is usually the case, will prevent the alignment of chains in a crystal lattice and hinder or prevent crystallization. Similarly, crosslinking restricts chain mobility and causes an increase in the apparent T_g. When the crosslinks are more frequent than every 40–50 main-chain atoms, the type of motion necessary to reach the rubbery state can never be achieved, and the polymer will degrade before reaching T_g.

REFERENCES

1. Tobolsky, A. V. *Properties and Structure of Polymers*, ch. II. John Wiley & Sons, Inc., New York, 1960.

2. Meares, P. The second-order transition of polyvinyl acetate. *Trans. Faraday Soc.*, **53,** no. 31, 1957.

3. Mandelkern, L.   The melting of crystalline polymers, *Rubber Chem. and Technol.*, **32,** p. 1392, 1959.

Section 2

Polymer
Synthesis

9
Polycondensation Reactions

9.1 INTRODUCTION

Polymerization reactions may be written generally as

$$x\text{-mer} + y\text{-mer} \rightarrow (x + y)\text{-mer}. \qquad (9\text{-}1)$$

In reactions which lead to condensation polymers, x and y may assume any value; i.e., chains of any size may react together as long as they are capped with the complementary functional groups. For this reason, these reactions are sometimes known as *chain-growth* polymerizations. In addition polymerization, although x may assume any value, y is confined to unity; i.e., a growing chain can react only with a monomer molecule and continue its growth. Thus, addition reactions are often termed *step-growth* polymerizations. This is probably the most fundamental classification of the two types of polymerization, because it avoids the difficulty caused by the formation of condensation polymers without the elimination of a small molecule —e.g., by ring scission, in which the small molecule has been eliminated previously in the formation of the cyclic monomer.

Regardless of the type of polymerization reaction, quantitative treatments are usually based on a common assumption— namely, that the reactivity of the functional group at a chain end is independent of the length of the chain. For example, in nylon 6-6

$$H\left[\overset{\overset{\displaystyle H}{|}}{N}\left(CH_2\right)_6\overset{\overset{\displaystyle H}{|}}{N}\overset{\overset{\displaystyle O}{\|}}{C}\left(CH_2\right)_4\overset{\overset{\displaystyle O}{\|}}{C}\right]_x OH$$

the rate constant for the reaction of the terminal amine and acid groups does not depend on x. Experimentally, this is an excellent assumption for x's greater than about five or six. Since most polymers must develop x's on the order of a hundred or so to be of practical value, it introduces little error.

With these basic concepts in mind, we proceed to a more detailed and quantitative treatment of polymerization reactions.

9.2 STATISTICS OF POLYCONDENSATION

Consider the two equivalent linear polycondensation reactions, assuming only difunctional monomers and stoichiometric equivalence.

$$\frac{x}{2}(\text{ARA}) + \frac{x}{2}(\text{BR'B}) \rightarrow \text{ARA} \left[\text{BR'B---ARA} \right]_{(x/2)-1} \text{BR'B}^* \quad (9\text{-}2a)$$

$$x\,(\text{ARB}) \qquad\qquad \rightarrow \text{ARB} \left[\text{ARB} \right]_{x-2} \text{ARB} \quad (9\text{-}2b)$$

It is important to note that x is used here to denote the *number of monomer residues*, or *structural units* in the chain, rather than repeating units.

Each of the polymer molecules above contains a total of x A groups:

>x-1 *reacted* A groups
>
>1 *unreacted* A group (on the end).

Let p = probability of finding a *reacted* A group (i.e., the *conversion* or *extent of reaction*),

   $(1 - p)$ = the probability of finding an *unreacted* A group.

     N = the total number of molecules present in the reaction mass (*of all sizes*)

    n_x = the number of molecules containing x A groups, both reacted and unreacted $\left(N = \sum_{x=1}^{\infty} n_x \right)$.

(*Some of the molecules formed in this type of reaction will be capped with two A groups. For each of these, however, there will be one capped by two B groups, so (9-2a) represents the *average* reaction.)

Now the total probability of finding a molecule with x A groups (reacted and unreacted) is equal to the mole or number fraction of those molecules present in the reaction mass, n_x/N. This in turn is equal to the probability of finding a molecule with $(x - 1)$ reacted A groups and one unreacted A group, and, since the total probability is the product of the individual probabilities,

$$n_x/N = p^{(x-1)}(1 - p) = \text{mole (number) fraction } x\text{-mer.} \quad (9\text{-}3)$$

This result gives the *distribution* of chain lengths in the re-action mass as a function of the conversion. The distribution results from the random nature of the reaction between chains of different length. The distribution is plotted in Figure 9.1

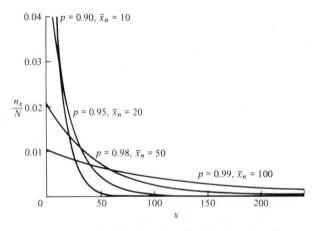

Figure 9.1.   Number (mole) fraction distributions for linear poly-condensations (Eq. 9-3).

for several conversions.   Note that the shorter chains are always more numerous; i.e., the longer the chain length, the fewer there are.

9.3 NUMBER-AVERAGE CHAIN LENGTHS

Since a distribution of chain lengths always arises in these condensation reactions, the average chain length is often of interest. If N_o = the original number of molecules present in the reaction mass, at the start of the reaction there are N_o unreacted A groups present, and at some time during the reaction there are N unreacted A groups, so N_o-N of the A groups have reacted. Thus

$$p = (N_o-N)/N_o = \text{fraction of reacted A groups present.} \quad (9\text{-}4)$$

Because the N_o original monomer molecules are distributed among the N molecules present in the reaction mass, the average number of monomer units per chain, $\bar{x}_n$, is given by

$$\bar{x}_n = N_o/N = \text{average number of monomer units per chain.} \quad (9\text{-}5)$$

When (9-4) and (9-5) are combined,

$$\bar{x}_n = 1/(1 - p) \quad \text{(Carothers' equation).} \quad (9\text{-}6)$$

This rather simple conclusion was reached by W. H. Carothers, the discoverer of nylon and one of the early founders of polymer science, in the 1930s, but it is often ignored to this day. Its importance becomes obvious when it is realized that typical linear polymers must have $\bar{x}_n$'s on the order of 100 to achieve useful mechanical properties. This requires a conversion of *at least* 99 per cent, if one assumes only difunctional monomers in perfect stoichiometric equivalence. Such high conversions are almost unheard of in most organic reactions, but they are necessary in most cases to achieve high molecular weight condensation polymers.

Example 1. Crud Chemicals is producing a linear polyester in a batch reactor by the condensation of a hydroxy acid

$$\text{HO—R—}\overset{\displaystyle O}{\overset{\|}{C}}\text{—OH.}$$

It has been proposed to follow the progress of the reaction by measuring the amount of water evolved from the reaction mass.

Assume pure monomer and assume that all the water of condensation can be removed and measured; then derive for them an equation relating $\bar{x}_n$ of the polymer to N_o, the moles of monomer charged, and M, the total moles of water evolved since the start of the reaction.

Solution. In this polycondensation reaction, each reacted

$$\text{—OH (or —}\overset{\displaystyle O}{\overset{\|}{C}}\text{—OH)}$$

group produces one molecule of water.

Therefore, the moles of reacted —OH (or $-\overset{\overset{\displaystyle O}{\displaystyle \|}}{C}$—OH) groups
is M. The probability of finding a reacted group (of either
kind) is

$$p = \frac{M}{N_o} = \frac{\text{moles reacted —OH groups}}{\text{total moles —OH groups}}.$$

If it is plugged into Carothers' equation, (9-6),

$$\bar{x}_n = \frac{1}{1 - \left(\dfrac{M}{N_o}\right)}.$$

If the restriction on perfect stoichiometric equivalence is
removed, similar (but more involved) reasoning leads to:

$$\bar{x}_n = \frac{1 + r}{2r(1 - p) + (1 - r)} = \frac{1 + r}{1 + r - 2rp} \qquad (9\text{-}7)$$

where r is the stoichiometric ratio of functional groups present,
N_A/N_B (A is taken to be the limiting reagent, so r is always less
than 1, and p represents the fraction of A groups reacted).

To examine the effect of stoichiometric imbalance, consider
the limiting case of complete conversion, $p = 1$, for which (9-7)
reduces to

$$\bar{x}_n = (1 + r)/(1 - r) \qquad \text{(for } p = 1). \qquad (9\text{-}8)$$

At $r = 1$, all the reactant molecules are combined in a single
molecule of essentially infinite molecular weight. Reducing
r to 0.99 cuts $\bar{x}_n$ to 199; to 0.95, all the way down to 39. Thus,
to reach high chain lengths, a close approach to stoichiometric
equivalence is required in addition to high conversions. Considering the usual industrial purity levels and the precision of
weighing techniques, this is not always easy to achieve. Fortunately, in the production of nylons from diacids and diamines, the monomers form an ionic salt of perfect 1:1 ratio
which is carefully purified before polymerization. Also, chain
length may sometimes be pushed up despite a stoichiometric
imbalance by an *ester interchange* reaction. For example, if a
polyester is formed from a diacid plus an excess of glycol, the
reaction will stop at a point when all the chains are capped by

—OH's. If the glycol is volatile enough to be driven off with the application of heat and vacuum (ethylene glycol, $R = CH_2—CH_2$, is about the only one which is volatile enough at temperatures below the degradation point), the chains may combine further through ester interchange with the elimination of glycol.

$$HO—R—O{\left[C(=O)—R'—C(=O)—O—R—O\right]}_x H \; +$$

$$H{\left[O—R—O—C(=O)—R'C(=O)\right]}_y O—R—OH \; \rightarrow$$

$$HO—R—O{\left[C(=O)—R'—C(=O)—O—R—O\right]}_{(x+y)} H \; + \; HOROH$$

On the other hand, introduction of an excess of one of the monomers or of some monofunctional material to reduce r deliberately provides a convenient means of limiting chain length.

If one of the monomers is at least trifunctional and if the proper stoichiometry is maintained, a conversion will be reached at which the reaction mass is crosslinked. At this *gel point*, the material becomes insoluble and infusible, so, if it is to be processed further, it must be removed from the reactor before the gel point. The statistics of gelation are treated in detail by Flory (*1*).

9.4 CHAIN LENGTHS ON A WEIGHT BASIS

The previous development of the distribution of chain lengths in a linear condensation polymer, although perfectly legitimate, is in some ways misleading, because it describes the *number* of molecules of a given chain length present and counts equally both monomer units and chains containing many hundreds of monomer units; i.e., each is one molecule. For example, in a mixture consisting of one monomer molecule and one 100-mer, the *number* or *mole* fraction monomer is $1/2$. Another way of looking at it is to inquire about the relative *weights* of the various chain lengths present. On this basis, the *weight fraction* monomer in the mixture is only $1/101$. By neglecting the weight of the small molecule split out in most condensation reactions, we can obtain the distribution of chain lengths in terms of weight fractions. If (w_x/W) = weight fraction of x-mer present = weight of x-mers/weight of batch, and if we let m = average molecular weight of a structural unit,

$$\text{weight of } x\text{-mer} = mxn_x \qquad (9\text{-}9)$$

$$\text{weight of batch} = mN_o \qquad (9\text{-}10)$$

$$\left(\frac{w_x}{W}\right) = xn_x/N_o. \qquad (9\text{-}11)$$

Combining (9-3), (9-4) and (9-11) and eliminating the N's gives

$$\frac{w_x}{W} = xp^{(x-1)}(1 - p)^2. \qquad (9\text{-}12)$$

This *weight fraction* distribution is shown in Figure 9.2. Here, although the monomer molecules are still the most numerous, their combined weight is an insignificant portion of the total weight. The peak is at $x = -1/\ln p \cong 1/(1 - p)$; i.e., the

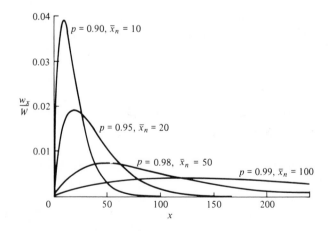

Figure 9.2.   Weight fraction distributions for linear polycondensations (Eq. 9-12).

x-mer present in greatest weight is *approximately* $\bar{x}_n$. The neglect of the molecule of condensation in this derivation can lead to significant errors at low x's where a molecule is split out. The exact solution is quite complex, and it has only recently been published (2).

The weight-average chain length, $\bar{x}_w$, can be obtained by inserting the number distribution, (9-3), into (6-19) and integrating, realizing that $\overline{M}_w = m\bar{x}_w$ and $M = mx$:

$$\bar{x}_w = \frac{\int_0^\infty n_x x^2 dx}{\int_0^\infty n_x x\, dx} = \frac{\int_0^\infty x^2 p^{(x-1)} dx}{\int_0^\infty x\, p^{(x-1)} dx} = \frac{1+p}{1-p} \cdot \quad (9\text{-}13)$$

(Evaluation of the above integrals is a good exercise for mathematical masochists.) Combining (9-13) and (9-6) gives an expression for the polydispersity index in terms of conversion

$$\frac{\bar{x}_w}{\bar{x}_n} = (1+p) \qquad (0 \le p \le 1). \qquad (9\text{-}14)$$

9.5 KINETICS OF POLYCONDENSATION

The kinetics of polycondensation is similar to that of ordinary condensation reactions. Since the average chain length is related to the extent of reaction (conversion), which in turn is given as a function of time by the kinetic expressions, $\bar{x}_n$ is directly related to the batch reaction time and can thus be controlled by stopping the reaction at the appropriate time.

Example 2. For a second-order polycondensation reaction (i.e., rate proportional to the concentrations of reactive A and B groups), obtain expressions for conversion and number-average chain length as a function of time for a stoichiometrically equivalent batch.

Solution. For a second-order reaction, assuming constant volume,

$$\frac{-d[A]}{dt} = k[A][B], \qquad (9\text{-}15)$$

where k is the reaction-rate constant and the brackets indicate concentrations. For $[A_o] = [B_o]$ (i.e., equivalent initial concentration), [A] = [B] at all times, and

$$-\frac{d[A]}{dt} = k[A]^2. \tag{9-16}$$

Separating variables and integrating between the limits when $t = 0$, $[A] = [A_o]$ and when $t = t$, $[A] = [A]$ gives

$$\frac{1}{[A]} - \frac{1}{[A_o]} = kt. \tag{9-17}$$

Since
$$p = \frac{[A_o] - [A]}{[A_o]}, \tag{9-4a}$$

a combination of (9-4a) and (9-17) and rearrangement gives

$$p = \frac{[A_o] kt}{1 + [A_o]kt}. \tag{9-18}$$

By combining (9-6) and (9-18),

$$\overline{x}_n = 1 + [A_o] kt; \tag{9-19}$$

i.e., the number-average chain length increases linearly with time.

Not all polycondensations are second order, as in the preceding example. Some polyesterifications, for example, are catalyzed by their own acid groups and are therefore first order in hydroxyl concentration, second order in acid, and third order overall. The rate may also be proportional to the concentration of an added catalyst (usually acids or bases), if used.

REFERENCES

1. Flory, P. J. *Principles of Polymer Chemistry*, Cornell University Press, Ithaca, N.Y., 1953.

2. Grethlein, H. E. Exact weight fraction distribution in linear condensation polymerization. *Ind. and Eng. Chem. Fundam.*, **8,** no. 206, 1969.

10
Free-Radical Addition Polymerization

10.1 INTRODUCTION

One of the most important types of addition polymerization is initiated by the action of *free radicals*, electrically neutral species with an *unshared electron*. In the developments to follow, a dot · will represent a single electron. The single bond, a pair of shared electrons, will be denoted by a double dot : or, where not necessary to indicate electronic configurations, by the usual —. A double bond, *two* shared electron pairs, is :: or =.

Free radicals for the initiation of addition polymerization are usually generated by the thermal decomposition of organic peroxides or azo compounds. Two common examples are benzoyl peroxide

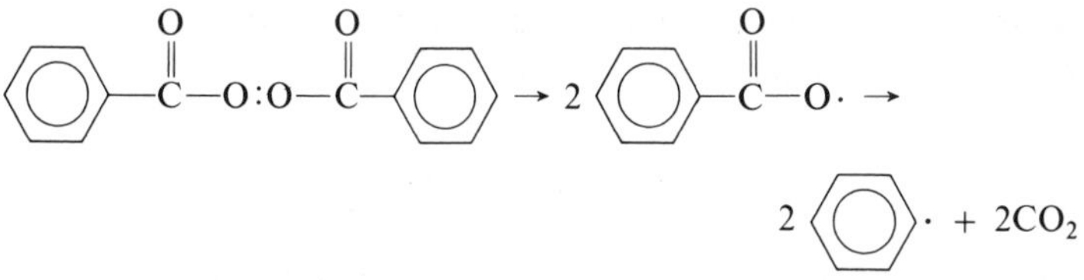

$$\langle \bigcirc \rangle - \overset{\overset{O}{\|}}{C} - O:O - \overset{\overset{O}{\|}}{C} - \langle \bigcirc \rangle \rightarrow 2 \langle \bigcirc \rangle - \overset{\overset{O}{\|}}{C} - O \cdot \rightarrow$$

$$2 \langle \bigcirc \rangle \cdot + 2CO_2$$

and azobisisobutyronitrile

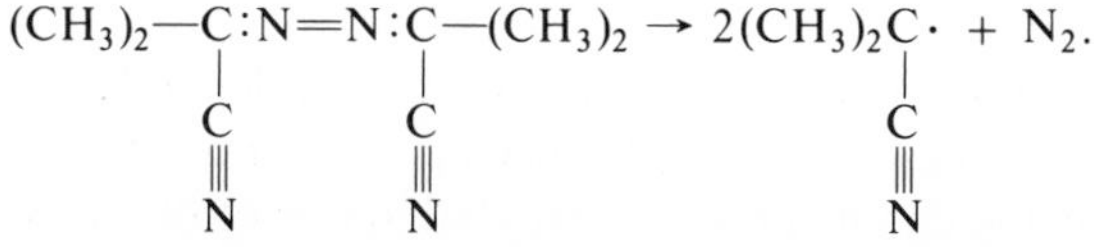

$$(CH_3)_2 - \underset{\underset{N}{\overset{\|}{C}}}{\overset{\|}{C}} : N = N : \underset{\underset{N}{\overset{\|}{C}}}{\overset{\|}{C}} - (CH_3)_2 \rightarrow 2(CH_3)_2 \underset{\underset{N}{\overset{\|}{C}}}{C} \cdot + N_2.$$

99

10.2 MECHANISM OF POLYMERIZATION

The initiator molecule, represented by I, decomposes by a first-order reaction with a rate constant k_d to give two free radicals, $R \cdot$

$$I \xrightarrow{k_d} 2R \cdot \quad \text{(decomposition)} \qquad (10\text{-}1)$$

The radical then adds a monomer unit by grabbing an electron from the electron-rich double bond, forming a single bond with the monomer but leaving an unshared electron at the other end.

$$
\begin{array}{c}
\quad\; H\;\; H \qquad\quad H\;H \\
\quad\; | \;\;\; | \qquad\qquad | \;\; | \\
R \cdot \; + \; C::C \;\rightarrow\; R:C:C\cdot \\
\quad\; | \;\;\; | \qquad\qquad | \;\; | \\
\quad\; H\;\; X \qquad\quad\; H\;X
\end{array}
$$

This may be abbreviated by

$$R \cdot \; + \; M \xrightarrow{k_a} M \cdot_1 \quad \text{(addition)}. \qquad (10\text{-}2)$$

Note that the product of the addition reaction is still a free radical; it proceeds to propagate the chain by adding another monomer unit

$$M \cdot_1 \; + \; M \xrightarrow{k_p} M \cdot_2$$

again maintaining the unshared electron at the chain end, which adds another monomer unit

$$M \cdot_2 \; + \; M \xrightarrow{k_p} M \cdot_3$$

and so on. In general, the propagation reaction is written as

$$M \cdot_x \; + \; M \xrightarrow{k_p} M \cdot_{(x+1)} \quad \text{(propagation)}. \qquad (10\text{-}3)$$

We have again assumed that reactivity is independent of chain length by using the same k_p for each propagation step.

A growing chain can be terminated in one of two ways:

$$
\text{~~}\underset{\underset{\text{H}}{|}}{\overset{\overset{\text{H}}{|}}{\text{C}}}-\underset{\underset{\text{X}}{|}}{\overset{\overset{\text{H}}{|}}{\text{C}}}\cdot \; + \; \cdot\underset{\underset{\text{X}}{|}}{\overset{\overset{\text{H}}{|}}{\text{C}}}-\underset{\underset{\text{H}}{|}}{\overset{\overset{\text{H}}{|}}{\text{C}}}\text{~~} \;\rightarrow\; \text{~~}\underset{\underset{\text{H}}{|}}{\overset{\overset{\text{H}}{|}}{\text{C}}}-\underset{\underset{\text{X}}{|}}{\overset{\overset{\text{H}}{|}}{\text{C}}}:\underset{\underset{\text{X}}{|}}{\overset{\overset{\text{H}}{|}}{\text{C}}}-\underset{\underset{\text{H}}{|}}{\overset{\overset{\text{H}}{|}}{\text{C}}}\text{~~}
$$

$$
\text{M}\cdot_x \;+\; \text{M}\cdot_y \xrightarrow{\;k_{tc}\;} \text{M}_{(x+y)} \quad \text{(combination)} \qquad \text{(10-4a)}
$$

or

$$
\text{~~}\underset{\underset{\text{H}}{|}}{\overset{\overset{\text{H}}{|}}{\text{C}}}-\underset{\underset{\text{X}}{|}}{\overset{\overset{\text{H}}{|}}{\text{C}}}\cdot \; + \; \cdot\underset{\underset{\text{X}}{|}}{\overset{\overset{\text{H}}{|}}{\text{C}}}-\underset{\underset{\text{H}}{|}}{\overset{\overset{\text{H}}{|}}{\text{C}}}\text{~~} \;\rightarrow\; \text{~~}\underset{\underset{\text{X}}{|}}{\overset{\overset{\text{H}}{|}}{\text{C}}}=\underset{}{\overset{\overset{\text{H}}{|}}{\text{C}}} \;+\; \text{H}:\underset{\underset{\text{X}}{|}}{\overset{\overset{\text{H}}{|}}{\text{C}}}-\underset{\underset{\text{H}}{|}}{\overset{\overset{\text{H}}{|}}{\text{C}}}\text{~~}
$$

$$
\text{M}\cdot_x \;+\; \text{M}\cdot_y \xrightarrow{\;k_{td}\;} \text{M}_x \;+\; \text{M}_y \quad \text{(disproportionation).} \quad \text{(10-4b)}
$$

The relative proportion of each depends on the particular polymer and the reaction temperature.

Example 1. A free radical initiator $R:R(R:R \rightarrow 2R\cdot)$ and a vinyl monomer $\underset{\underset{\text{H}}{|}}{\overset{\overset{\text{H}}{|}}{\text{C}}}=\underset{\underset{\text{X}}{|}}{\overset{\overset{\text{H}}{|}}{\text{C}}}$ are polymerized in a homogeneous reaction mass. The resulting molecules all have the structures

$$
\text{A}\!-\!\!\left[\underset{\underset{\text{H}}{|}}{\overset{\overset{\text{H}}{|}}{\text{C}}}-\underset{\underset{\text{X}}{|}}{\overset{\overset{\text{H}}{|}}{\text{C}}}\right]\!\!-\!\text{B} \quad\text{or}\quad \text{D}\!-\!\!\left[\underset{\underset{\text{H}}{|}}{\overset{\overset{\text{H}}{|}}{\text{C}}}-\underset{\underset{\text{X}}{|}}{\overset{\overset{\text{H}}{|}}{\text{C}}}\right]_{\!x}\!\!\left[\underset{\underset{\text{X}}{|}}{\overset{\overset{\text{H}}{|}}{\text{C}}}-\underset{\underset{\text{H}}{|}}{\overset{\overset{\text{H}}{|}}{\text{C}}}\right]_{\!y}\!\!\!-\!\text{E}.
$$

$$
\text{I} \qquad\qquad\qquad\qquad \text{II}
$$

Tabulate all possible combinations of (a) groups A and B and (b) of groups D and E. Interchanging A and B or D and E are not considered separate combinations.

Solution. Molecules of type I are terminated by disproportionation; those of type II, by combination. Therefore

a. $\dfrac{\text{A}}{\text{R}}\qquad \dfrac{\text{B}}{\text{H}}$ b. $\dfrac{\text{D}}{\text{R}}\qquad \dfrac{\text{E}}{\text{R}}$

$$
\text{R} \qquad -\underset{}{\overset{\overset{\text{H}}{|}}{\text{C}}}=\underset{\underset{\text{H}}{|}}{\overset{\overset{\text{H}}{|}}{\text{C}}}
$$

10.3 KINETICS OF POLYMERIZATION

In practice, not all the radicals generated in reaction (10-1) actually initiate chain growth as in reaction (10-2). Some recombine or are used up by side reactions. By letting f = fraction of radicals generated which actually do initiate chain growth, the rate of formation of growing chain radicals through reactions (10-1) and (10-2) is

$$r_i = \left(\frac{d[M\cdot]}{dt}\right)_i = 2fk_d[I] \quad \text{(rate of initiation).} \quad (10\text{-}5)$$

Although it is rarely mentioned explicitly, this expression is based on the generally valid assumption that the rate of reaction (10-2) is much greater than that of (10-1)—i.e., that the initiator decomposition is *rate controlling*. Thus, as soon as an initiator radical is formed, it grabs a monomer molecule, starting chain growth, so k_a does not appear in the expression.

According to reaction (10-3), the rate of monomer removal in the propagation step is

$$r_p = \left(\frac{d[M]}{dt}\right)_p = -k[M][M\cdot] \quad \text{(rate of propagation)} \quad (10\text{-}6)$$

where $[M\cdot]$ is the total concentration of growing chain radicals, regardless of length.

The rate of removal of chain radicals is the sum of the rates of the two termination reactions. Since both are second order,

$$r_t = \frac{d[M\cdot]}{dt} = -2k_t[M\cdot]^2 \quad \text{(rate of termination)} \quad (10\text{-}7)$$

where

$$k_t = (k_{tc} + k_{td}). \quad (10\text{-}8)$$

Unfortunately, the unknown quantity $[M\cdot]$ is present in the equations. It can be removed by making the standard kinetic assumption of a steady-state concentration of a transient species—in this case, the chain radicals. In order that $[M\cdot]$ remain constant, chain radicals must be generated at the same rate at which they are removed, $r_i = -r_t$, or

$$2fk_d[I] = 2k_t[M\cdot]^2. \quad (10\text{-}9)$$

This gives the chain-radical concentration

$$[M\cdot] = \left(\frac{fk_d[I]}{k_t}\right)^{1/2} \qquad (10\text{-}10)$$

which, when inserted into (10-6), provides an expression for the rate of monomer removal in the propagation reaction:

$$r_p = \left(\frac{d[M]}{dt}\right)_p = -k_p\left(\frac{fk_d[I]}{k_t}\right)^{1/2}[M]. \qquad (10\text{-}11)$$

Actually, monomer is consumed in the addition step (10-2) also, but, for long chains, the amount is insignificant so that (10-11) is the overall rate of polymerization—the rate at which monomer is converted to polymer.

Equation (10-11) is the classical rate expression for a *homogeneous*, free-radical polymerization. It has been well substantiated in many important cases, particularly in fairly dilute solutions. In certain cases of industrial importance, however, significant deviations are observed. The nature of these will be discussed in chapter 14 on polymerization practice.

Of greater interest to engineers is the integrated form of (10-11), giving the conversion $([M_o] - [M])/[M_o]$ as a function of time. A form occasionally used by chemists in studying the initial stages of an isothermal, batch reaction is obtained by assuming constant initiator concentration $[I_o]$.

$$\ln\frac{[M]}{[M_o]} = -k_p\left(\frac{fk_d[I_o]}{k_t}\right)^{1/2}t \qquad (10\text{-}12)$$

The assumption of constant $[I_o]$ is not likely to be realistic in cases of industrial importance beyond a few percent conversion, however. By including the first-order decay of the initiator, starting at time $t = 0$ with $[I] = [I_o]$

$$\frac{d[I]}{dt} = -k_d[I] \qquad (10\text{-}13\text{a})$$

$$[I] = [I_o]e^{-k_d t} \qquad (10\text{-}13\text{b})$$

and inserting (10-13b) into (10-11) and integrating, a more realistic expression for monomer concentration versus time is obtained:

$$\ln \frac{[M]}{[M_o]} = \frac{2k_p}{k_d} \left(\frac{fk_d[I_o]}{k_t} \right)^{1/2} [e^{-(k_d t/2)} - 1]. \quad (10\text{-}14)$$

This expression has some interesting and important implications when compared with (10-12). Setting $t = \infty$ reveals that there is a maximum attainable conversion which depends on $[I_o]$:

$$\text{Max. Conv.} = \left(1 - \frac{[M]}{[M_o]} \right)_{max} = 1 - \exp \left\{ -\frac{2}{k_d} \left(\frac{k_p^2}{k_t} fk_d[I_o] \right)^{1/2} \right\}.$$

$$(10\text{-}15)$$

This dead-stop situation is basically a matter of the initiator being used before the monomer, but it is not revealed by (10-12) which always predicts complete conversion at long enough times. Thus, the use of (10-12) instead of (10-14) can result in considerable error at high conversions, although they approach each other at low conversions.

Example 2. For the polymerization of pure styrene with azobisisobutyronitrile at 60C, in an isothermal, batch reactor,

a. compare equations (10-12) and (10-14) by plotting conversion $1 - [M]/[M_o]$ versus time (use the data given below), and

b. determine the minimum $[I_o]$ needed to achieve 90 percent conversion. Notice that, according to (10-12), there is *no* minimum.

Data: $k_p^2/k_t = 1.18 \times 10^{-3}$ liter/mole, sec

ρ styrene $= 0.907$ g/cc

$k_d = 0.96 \times 10^{-5} \text{ sec}^{-1}$

$[I_o] = 0.05$ mole/liter

$f = 1.0$

Solution.

a. The molecular weight of styrene is 104. Therefore,

$$[M_o] = 0.907 \frac{g}{cc} \times \frac{1000 \text{ cc}}{\text{liter}} \times \frac{1 \text{ g mol}}{104 \text{ g}} = 8.7 \frac{\text{g moles}}{\text{liter}}$$

Plugging this and the data into (10-12) and (10-14) results in the plots shown in Figure 10.1. The maximum conversion under these conditions is 99.3 percent (10-14).

b. Setting $(1 - [M]/[M_o])_{max} = 0.90$ in (10-15) and solving gives an $[I_o]_{min} = 0.0108$ moles/liter.

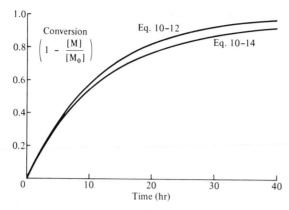

Figure 10.1. Conversion versus time for isothermal, free radical batch polymerization (data of example 1).

Side reactions sometimes cause deviations from the classical kinetics. Agents which cause these reactions are generally categorized as *inhibitors* or *retarders*. An inhibitor delays the start of the reaction, but, once begun, it proceeds at the normal rate. Vinyl monomers are normally shipped with a few parts per million inhibitor to prevent polymerization in transit. A retarder slows down the reaction rate. Some chemicals combine both effects. These are illustrated in Figure 10.2. Oxygen is an effective inhibitor for the free-radical polymerization of vinyl monomers, so the reactions are normally carried out under a blanket of nitrogen.

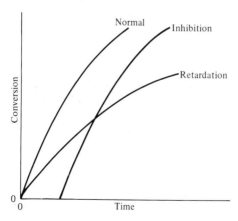

Figure 10.2. Inhibition and retardation.

10.4 AVERAGE CHAIN LENGTHS

As in condensation polymerization, a distribution of chain lengths is always obtained in a free-radical addition polymerization because of the inherently random nature of the termination reaction with regard to chain length. Theoretical calculation of the distribution is more complex in this case, however, and the results are less likely to be valid due to a wide variety of side reactions which are possible. The subject is treated in the literature (*1*) but will not be considered here.

Fairly straightforward expressions for the number-average chain length may be derived. These are usually couched in terms of the *kinetic chain length*, ν, which is the rate of monomer addition to growing chains over the rate at which chains are started by initiator radicals—i.e., the average number of monomer units per growing chain radical at a particular instant. It thus expresses the efficiency of the initiator radicals in polymerizing monomer.

$$\nu = \frac{r_p}{r_i} = \frac{k_p[M]}{2(fk_dk_t)^{1/2}[I]^{1/2}} \qquad (10\text{-}16)$$

If the growing chains terminate exclusively by disproportionation, they undergo no change in length in the process, but, if combination is the exclusive mode of termination, the growing chains, on the average, double in length upon termination. Therefore,

$$\overline{x}_n = \nu \qquad \text{(termination by disproportionation)} \qquad (10\text{-}17a)$$

$$\overline{x}_n = 2\nu \qquad \text{(termination by combination).} \qquad (10\text{-}17b)$$

The average chain length may be expressed more generally in terms of a quantity ξ (ξ = rate of dead chain formation/rate of termination reactions), the average number of dead chains produced per termination. Since each disproportionation reaction produces *two* dead chains and each combination reaction *one*,

$$\text{rate of dead chain formation} = (2k_{td} + k_{tc})[M\cdot]^2 \qquad (10\text{-}18)$$

$$\text{rate of termination reactions} = (k_{tc} + k_{td})[M\cdot]^2 \qquad (10\text{-}19)$$

$$\xi = \frac{k_{tc} + 2k_{td}}{k_{tc} + k_{td}} = \frac{k_{tc} + 2k_{td}}{k_t}. \qquad (10\text{-}20)$$

The instantaneous number-average chain length is the rate of addition of monomer units to all chains (r_p) over the rate of dead chain formation:

$$\bar{x}_n = \frac{k_p([M\cdot][M])}{(2k_{td} + k_{tc})[M\cdot]^2} = \frac{k_p[M]}{\xi(fk_dk_t)^{1/2}[I]^{1/2}}. \qquad (10\text{-}21)$$

When $k_{td} \gg k_{tc}$, $\xi = 2$ and when $k_{tc} \gg k_{td}$, $\xi = 1$, duplicating the previous result, but (10-21) also handles various degrees of mixed termination. This is particularly important when considering nonisothermal reactions where ξ may change appreciably with temperature.

Keeping in mind that $\bar{x}_n$ is one of the most important factors in determining certain mechanical properties of polymers, what is the significance of (10-21)? A growing chain may react with another growing chain and terminate, or it may add another monomer unit and continue its growth. The more monomer molecules in the vicinity of the chain radical, the higher the probability of another monomer addition—hence, the proportionality to [M]. On the other hand, the more initiator radicals there are present competing for the available monomer, the shorter the chains will be (on the average), causing the inverse proportionality to the square root of [I]. Equation (10-11) shows that the rate of polymerization can be increased (always a desirable economic goal) by increasing both [M] and [I], the former being more efficient than the latter. However, there is an upper limit to [M] set by the density of the pure monomer at the reaction conditions. So it is often tempting to increase [I] to attain higher rates. However, according to (10-21), this unavoidably lowers $\bar{x}_n$; i.e., you can't have your cake and eat it, too.

10.5 INSTANTANEOUS AND OVERALL AVERAGE CHAIN LENGTHS

Both the rate of polymerization and average chain length as given by (10-11) and (10-21) are instantaneous quantities—

functions of temperature (through the temperature dependence of the rate constants), [M] and [I], all of which may be varying with time and position in a reactor.

Example 3. Show how both the *instantaneous* number-average chain length, $\bar{x}_n(t)$, and the *overall* number-average chain length, $<\bar{x}_n>(t)$, may be obtained as functions of time from the kinetic expressions for an isothermal, homogeneous free-radical addition polymerization in a batch reactor.

Solution. $\bar{x}_n(t)$ is given by (10-21) with a knowledge of $[M](t)$ and $[I](t)$. $[M](t)$ is calculated from (10-14) and $[I](t)$ from (10-13b). The overall average at any time is simply the moles of monomer polymerized (in polymer chains) up to that time divided by the moles of chains formed to that time

$$\text{moles of monomer in chains} = [M_o] - [M](t)$$

$$\text{moles of chains formed} = f\xi([I_o] - [I](t)).$$

(For termination by disproportionation, each initiator molecule results in two chains; for combination, it results in one; hence, the ξ.) Therefore,

$$<\bar{x}_n>(t) = \frac{[M_o] - [M](t)}{f\xi([I_o] - [I](t))}.$$

The quantities $[M](t)$ and $[I](t)$ are obtained as above. Note that the preceding expression is indeterminate at $t = 0$, but $<\bar{x}_n>(0) = \bar{x}_n(0)$; both are given by (10-21) with $[M] = [M_o]$ and $[I] = [I_o]$. The quantity $<\bar{x}_n>(t)$ is what would be measured by sampling the reactor contents.

10.6 CHAIN TRANSFER

In practice, another type of reaction often occurs in free-radical addition polymerizations. These *chain-transfer* reactions kill a growing chain radical and start a new one in its place

$$R':H + M\cdot_x \xrightarrow{k_{tr}} M_x + R'\cdot \qquad (10\text{-}22a)$$

$$R'\cdot + M \xrightarrow{k_a'} M_1\cdot \quad \text{etc.} \qquad (10\text{-}22b)$$

Thus, chain transfer results in shorter chains, and, if reactions (10-22) are not too frequent compared to the propagation reaction and don't have very low rate constants, it will not change the overall rate of polymerization appreciably.

The compound $R':H$ is known as a *chain-transfer agent.* Under appropriate conditions, almost anything in the reaction mass may act as a chain-transfer agent, including initiator, monomer, solvent, and dead polymer.

Example 4. Show, by using dots to represent the electrons involved, how chain transfer to dead polymer leads to branching in polyethylene.

Solution.

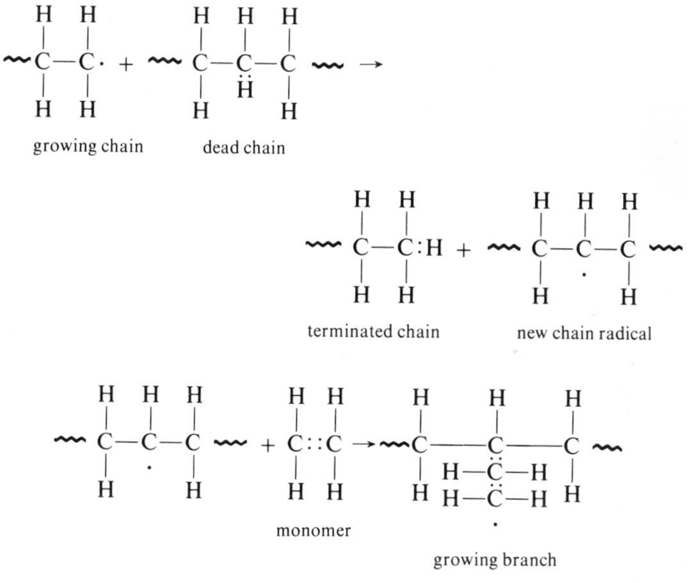

growing chain   dead chain

terminated chain   new chain radical

monomer

growing branch

Most frequently, mercaptans ($R'S:H$) are added to the reaction mass as effective chain-transfer agents to lower the average chain length.

Example 5. A mercaptan, $R'S:H$, is added to the reaction in example 1. Tabulate the *additional* possibilities for groups A and B and D and E.

Solution. Chains may now be terminated by the chain-transfer agent, and new ones started by it, giving rise to the additional possibilities

a.

A	B
R'S	H	(started by c-t, terminated by c-t or disp)

$$
\begin{array}{cc}
 & \begin{array}{cc} H & H \\ | & | \end{array} \\
R'S & -C=C \\
 & | \\
 & X
\end{array}
$$
(started by c-t, terminated by disp)

b.

D	E
R'S	R'S	(both ends started by c-t)
R	R'S	(one end started by initiator, other by c-t)

The rate of dead chain formation by chain transfer

$$
r_{tr} = \left(\frac{d[M\cdot]}{dt}\right)_{tr} = -k_{tr}[R':H][M\cdot] \tag{10-23}
$$

must be added to the denominator of (10-21) to give the total rate of dead chain formation

$$
\bar{x}_n = \frac{k_p[M\cdot][M]}{(2k_{td} + k_{tc})[M\cdot]^2 + k_{tr}[R':H][M\cdot]}. \tag{10-24}
$$

Using (10-10) and (10-20) gives

$$
\bar{x}_n = \frac{k_p[M]}{\xi(fk_dk_t[I])^{1/2} + k_{tr}[R':H]}. \tag{10-25}
$$

By taking the reciprocal of (10-25),

$$
\frac{1}{\bar{x}_n} = \frac{1}{(\bar{x}_n)_o} + C\frac{[R':H]}{[M]} \tag{10-26}
$$

where C is the *chain-transfer constant* $= k_{tr}/k_p$ and $(\bar{x}_n)_o$ is simply the average chain length in the absence of transfer (10-21). Thus, a plot of $1/\bar{x}_n$ versus $[R':H]/[M]$ is linear with a slope of C and an intercept of $1/(\bar{x}_n)_o$.

Example 6. Prove that the use of a chain-transfer agent with C = 1 will maintain the ratio $[R':H]/[M]$ constant in a batch reaction.

Solution.

$$
\frac{r_{tr}}{r_p} = \frac{k_{tr}[R':H][M\cdot]}{k_p[M][M\cdot]} = C\frac{[R':H]}{[M]} = \frac{-d[R':H]/dt}{-d[M]/dt}
$$

cancelling dt's, separating variables, and integrating from $t = 0$ gives

$$\int_{[R':H]_o}^{[R':H]} \frac{d[R':H]}{[R':H]} = C \int_{[M]_o}^{[M]} \frac{d[M]}{[M]} \quad \text{or} \quad \frac{[R':H]}{[R':H]_o} = \left(\frac{[M]}{[M]_o}\right)^C .$$

Therefore, only if $C = 1$ will $[R':H]/[M] = [R':H]_o/[M]_o =$ constant. C's greater than one use up chain transfer agent too quickly, and lower ones cause it to have little effect until high conversions are reached.

Note, however, that a $C = 1$ does not guarantee that $(\bar{x}_n)_o$ will remain constant, so, in general, $\bar{x}_n$ will still vary. Adjusting the rate of addition of chain-transfer agent to maintain $\bar{x}_n$ constant has been discussed for batch (2) and continuous (3) reactors.

10.7 COMPARISON OF CONDENSATION AND FREE-RADICAL ADDITION POLYMERIZATION

Before pushing on, it is worthwhile to review some of the differences between condensation and free-radical addition polymerization. In both cases, the inherently random nature of the reactions on a molecular scale gives rise to a distribution of chain lengths, making it necessary to consider an average chain length, $\bar{x}_n$. In condensation polymerization, a homogeneous reaction mass consists of a continuous distribution of chain lengths from monomer on up to extremely long chains. As the reaction proceeds, shorter chains hook up to form longer chains, continually increasing $\bar{x}_n$ over the reaction period (typically on the order of hours). As given by (9-7), $\bar{x}_n$ describes the entire contents of the reaction mass (excluding inerts) and is a function of the extent of reaction only.

In a homogeneous free-radical addition polymerization, the time required for the initiation, growth and termination of a chain is small compared to the reaction period (seconds versus hours). A chain formed early in the reaction merely sits around inertly as other chains are formed from the monomer. At any instant, then, the reaction mass consists of monomer and a distribution of high polymer chains (excluding inerts).

Equation (10-25) gives the average chain length of the *polymer* being formed *at a particular* instant in terms of T, [M], and [I] *at that instant.* Since T, [M] and [I] all may vary with time in the reactor, so may $\bar{x}_n$, further broadening the distribution. Thus, the actual polymer content of the reactor at any time must be characterized by an overall average $<\bar{x}_n>$ of the polymer formed up to that time. To complicate things further, T, [M], and [I] may vary with location in a reactor. These considerations play an important role in polymerization reactor design, but a quantitative treatment is beyond the scope of this discussion.

Example 7. Which type of isothermal reactor would produce the narrowest possible distribution of chain lengths in a free-radical addition polymerization: continuous stirred tank (backmix), batch (assume perfect stirring in each of the previous cases), a plug-flow tubular, or laminar-flow tubular?

Solution. Only in an ideal continuous stirred tank reactor are [M] and [I], and therefore $\bar{x}_n$, constant, giving the narrowest possible distribution, due only to the microscopically random nature of the reaction. In a batch reactor, [M] and [I] vary with time and, in tubular reactors, with position, thus causing a shift in $\bar{x}_n$ with conversion and broadening the distribution.

REFERENCES

1. Bamford, C. H., et. al. *The Kinetics of Vinyl Polymerization by Radical Mechanisms.* Academic Press, New York, 1958.

2. Hoffman, B. F., S. Schreiber and G. Rosen. Batch Polymerization. Narrowing molecular weight distribution. *Ind. and Eng. Chem.,* **56,** no. 5, p. 51, 1964.

3. Kenat, T., R. I. Kermode and S. L. Rosen. The continuous synthesis of addition polymers. *J. Appl. Polym. Sci.,* **13,** p. 1353, 1969.

11
Non-Radical Addition Polymerization

11.1 INTRODUCTION

Many unsaturated monomers can be made to undergo addition polymerization by other than free-radical mechanisms. These nonradical addition reactions are important because they have the possibility of sterically directing the addition of monomer to the growing chain, thereby producing stereoregular polymers.

11.2 CATIONIC POLYMERIZATION (1)

Strong Lewis acids (i.e., electron acceptors) are often capable of initiating the addition polymerization of monomers with electron-rich double bonds. Cationic catalysts are most commonly metal trihalides such as $AlCl_3$ or BF_3. These compounds, although electrically neutral, are two electrons short of having a complete valence shell of eight electrons. They were found to require traces of a cocatalyst, usually water, to initiate polymerization, first by grabbing a pair of electrons from the cocatalyst.

$$F:\overset{\text{F}}{\underset{\text{F}}{B}} + :\overset{\text{H}}{\underset{\text{H}}{O}}: \rightarrow F:\overset{\text{F}}{\underset{\text{F}}{B}}:\overset{\text{H}}{\underset{\text{H}}{O}}: \rightarrow \left[F:\overset{\text{F}}{\underset{\text{F}}{B}}:\overset{..}{\underset{..}{O}}:H\right]^{-} + [H]^{+}$$

The leftover proton is thought to be the actual initiating species, abstracting a pair of electrons from the monomer and leaving a cationic chain end which reacts with additional monomer molecules.

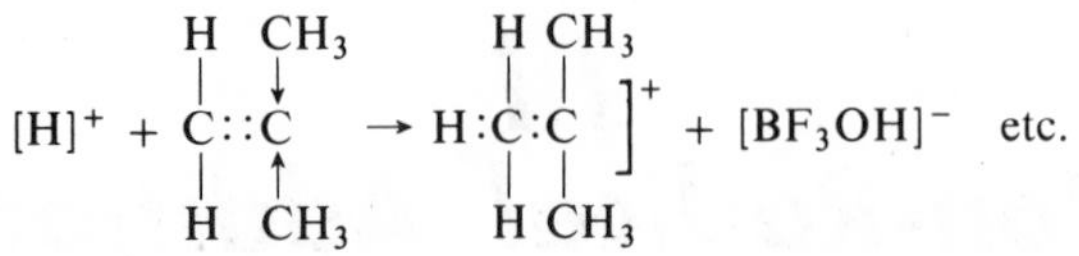

$$[H]^+ + \begin{matrix} H & CH_3 \\ | & | \\ C::C \\ | & | \\ H & CH_3 \end{matrix} \rightarrow \begin{bmatrix} H & CH_3 \\ | & | \\ H:C:C \\ | & | \\ H & CH_3 \end{bmatrix}^+ + [BF_3OH]^- \quad etc.$$

gegen or counter ion

An important point here is that the *gegen* or *counter* ion is electrostatically held near the growing chain end and so can exert a steric influence on the addition of monomer units. Termination is thought to occur by a disproportionation-like reaction which regenerates the catalyst complex. The complex, therefore, is a true catalyst, unlike free-radical initiators.

$$\begin{bmatrix} H & CH_3 \\ | & | \\ \sim\sim C-C \\ | & | \\ H & CH_3 \end{bmatrix}^+ + [BF_3OH]^- \rightarrow \begin{matrix} H & CH_3 \\ | & | \\ \sim\sim C-C \\ | & \| \\ H & CH_2 \end{matrix} + BF_3 \cdot H_2O$$

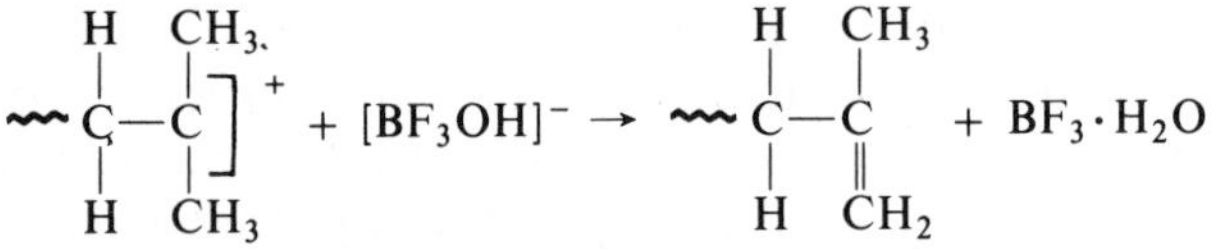

The kinetics of these reactions is not well understood, but they proceed very rapidly at extremely low temperatures. For example, the polymerization of isobutylene illustrated above is carried out commercially at $-150F$. The average chain length increases as the temperature is lowered.

Cationic initiation is successful only with monomers like isobutylene having electron-rich substituents adjacent to the double bond, such as

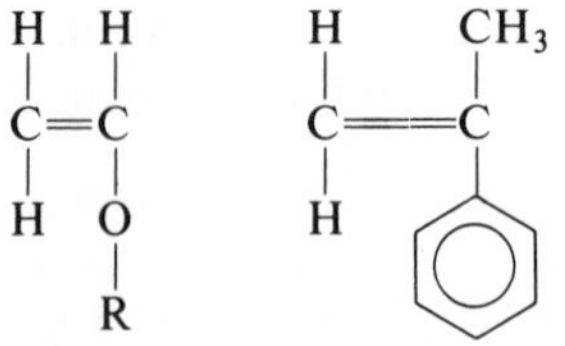

$$\begin{matrix} H & H \\ | & | \\ C=C \\ | & | \\ H & O \\ & | \\ & R \end{matrix} \qquad \begin{matrix} H & CH_3 \\ | & | \\ C==C \\ | & \\ H & \bigcirc \end{matrix}$$

alkyl vinyl ethers   α-methyl styrene

None of these monomers can be polymerized to high molecular weight with free-radical initiators.

11.3 ANIONIC POLYMERIZATION

Addition polymerization may also be initiated by anions. A wide variety of anionic initiators has been investigated, but the use of organic alkali-metal salts has recently begun to achieve great commercial importance. The mechanism of these reactions is illustrated below for the polymerization of styrene with *n*-butyllithium.

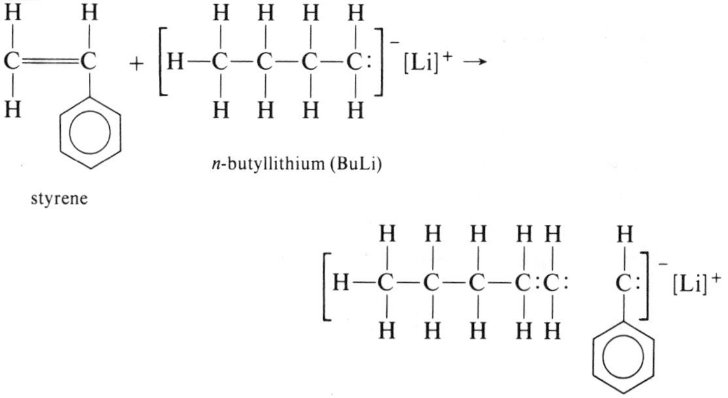

n-butyllithium (BuLi)

styrene

The anionic chain end then propagates the chain by adding another monomer molecule. Again, the gegen ion can sterically influence the reaction.

There are several very interesting aspects to these reactions. First, unlike free-radical reactions, the relative rates of propagation and initiation may vary widely, depending on the nature of the monomer, gegen ion and the solvent. In general, the propagation reaction proceeds more rapidly in polar solvents than in nonpolar solvents as the propagating chain end is more highly ionized. For many important cases, however, the rate of initiation is much greater than that of propagation, $r_i \gg r_p$ (just the reverse of free-radical addition). This means that the initiator (in this case, it is an initiator rather than a catalyst) starts chains growing immediately. Second, in an inert medium, *there is no termination step*. The chains continue to grow until the monomer supply is exhausted. The ionic chain end is perfectly stable, and the growth of the chains can be resumed by the addition of more monomer. For this reason, these materials have been aptly termed *"living" polymers* by

Professor Swarc. The presence of impurities such as water or alcohols can quickly kill (terminate) them, however.

$$\underset{\overset{\displaystyle |}{H}}{\overset{\overset{\displaystyle H}{|}}{\sim\sim C}} - \overset{\overset{\displaystyle H}{|}}{\underset{\bigcirc}{C:}}\Big]^{-}[Li]^{+} + H_2O \rightarrow \sim\sim \overset{\overset{\displaystyle H}{|}}{\underset{\overset{\displaystyle |}{H}}{C}} - \overset{\overset{\displaystyle H}{|}}{\underset{\bigcirc}{C}} :H + LiOH$$

Block Copolymerization　If, after the initial monomer charge is exhausted, a second monomer is introduced, the chains resume propagation with the second monomer, neatly giving a block copolymer. Monomers can be alternated as desired to give AB, ABA or other more complicated block structures, conceivably even including three or more monomers.

Monodisperse Polymers　Since all chains are started nearly simultaneously and the growing chains compete on an even basis for the available monomer, all the chains will be essentially of the same length (subject to some minor statistical variation). This is the only known technique for synthesizing nearly monodisperse polymers. (There is the important practical limitation of not being able to add and mix the initiator instantaneously and uniformly. This always causes some spread of chain length in practice.) These unique abilities are rapidly increasing the commercial importance of anionic polymerization.

11.4 ANIONIC KINETICS (2, 3)

A general description of the kinetics of anionic polymerization is complicated by the associations which can occur, particularly in nonpolar (hydrocarbon) solvents. Some useful approximations may be obtained, however. The rate of polymerization will be proportional to the product of the monomer concentration and the concentration of growing chain ends. Under conditions of negligible association (as in tetrahydrofuran solvent, for example, or hydrocarbons at BuLi concentrations less than 10^{-4} molar), each initiator molecule will start a growing chain, and, in the absence of terminating impurities, the number of growing chain ends will always equal the number of initiator molecules added, and, if propagation is rate-controlling,

$$r_p = \frac{-d[M]}{dt} = k_p[M][I_o]. \qquad (11\text{-}1)$$

In BuLi polymerizations at high concentrations in nonpolar solvents, the chain ends are present almost exclusively as inactive dimers, which dissociate slightly according to the equilibrium

$$(BuM_x^-Li^+)_2 \xrightarrow{K} 2BuM_x^-Li^+ \qquad (11\text{-}2)$$

where $K = [BuM_x^-Li^+]^2/[(BuM_x^-Li^+)_2] \ll 1$.

The concentration of *active* chain ends is then

$$[BuM_x^-Li^+] = K^{1/2}[BuM_x^-Li^+)_2]^{1/2}. \qquad (11\text{-}3)$$

Now it takes two initiator molecules to make one inactive chain dimer, so

$$[(BuM_x^-Li^+)_2] = \frac{[BuLi]}{2} = \frac{[I_o]}{2}. \qquad (11\text{-}4)$$

The rate of polymerization then becomes

$$r_p = \frac{-d[M]}{dt} = k_p K^{1/2}\left(\frac{[I_o]}{2}\right)^{1/2} \qquad (11\text{-}5)$$

The low value of K, reflecting the presence of most chain ends in the inactive associated state, gives rise to the low rates of polymerization in nonpolar solvents. At very high concentrations, association may be even greater and the rate essentially independent of $[I_o]$.

Since the polymerized monomer is split up evenly among the chains present

$$\bar{x}_n = \frac{\text{moles monomer polymerized}}{\text{moles chains present}} = \frac{[M_o] - [M]}{[I_o]}. \qquad (11\text{-}6)$$

As in the case of condensation polymerization, the chain length increases with conversion and is a function of conversion only in a homogeneous reaction mass. The reaction mass consists only of monomer molecules and polymer chains of essentially a single length. Equation (11-6) characterizes the polymer present.

Example 1. Consider the anionic batch polymerization of styrene in tetrahydrofuran solution, $[M_o] = 1.0$ g moles/liter, with *n*-butyllithium initiator. By assuming that the propagation step is rate controlling and that mixing is perfect and instantaneous,

a. for an isothermal reaction (constant k_p), obtain an expression relating $\bar{x}_n$ to time;

b. the reaction is started with $[I_o] = 1 \times 10^{-3}$ g moles/liter. At 50 percent conversion, water is added instantaneously to a concentration of $[H_2O] = 0.5 \times 10^{-3}$ g moles/liter. At 100 percent conversion, what chain lengths will be present in the reaction mass?

c. For case (b), calculate $\bar{x}_n$ at 100 percent conversion.

Solution.

a. There will be no association in this polar solvent. Separating variables and integrating (11-1) gives

$$[M] = [M_o]e^{-k_p[I_o]t} \qquad (11\text{-}7)$$

After plugging it into (11-6),

$$x = \frac{[M_o] - [M_o]e^{-k_p[I_o]t}}{[I_o]} = \frac{[M_o]}{[I_o]}(1 - e^{-k_p[I_o]t}). \quad (11\text{-}8)$$

(As the polymer will be essentially monodisperse, no bar and subscript are required on x.)

b. The initiator starts 1×10^{-3} g moles/liter of chains growing. At 50 percent conversion, the water terminates half the chains. Thus, $(1.0/4)$ g moles/liter of monomer is present in the 0.5×10^{-3} g moles/liter of terminated chains.

$$x \text{ (terminated chains)} = \frac{1.0}{4(0.5 \times 10^{-3})} = 5 \times 10^2$$

The remaining $1.0(3/4)$ g moles/liter monomer continues to grow in 0.5×10^{-3} g moles/liter of unterminated chains.

$$x \text{ (unterminated chains)} = 1.0\left(\frac{3}{4}\right)\frac{1}{0.5 \times 10^{-3}} = 15 \times 10^2$$

c. Since all the initiator starts chains, and [M] = 0 at 100 percent conversion,

$$\bar{x}_n = \frac{[M_o] - [M]}{[I_o]} = \frac{1.0 - 0}{1 \times 10^{-3}} = 10 \times 10^2$$

or

$$\bar{x}_n = \frac{\Sigma n_x x}{\Sigma n_x} = \frac{(0.5 \times 10^{-3})(5 \times 10^2) + (0.5 \times 10^{-3})(15 \times 10^2)}{0.5 \times 10^{-3} + 0.5 \times 10^{-3}}$$

$$= 10 \times 10^2.$$

Example 2. Which type of isothermal reactor will produce the narrowest possible distribution of chain lengths in an anionic polymerization, batch, continuous stirred tank (backmix) (assume both perfectly stirred), plug-flow tubular or laminar flow tubular?

Solution. As indicated by (11-8) in example 1, the chain length depends only on how long a chain is allowed to grow. In a batch reactor and an *ideal* plug-flow reactor, all chains react for the same length of time; hence, the product will be essentially monodisperse. In a CSTR and a laminar-flow tubular reactor, the residence time of chains in the reactor varies, causing a spread in the distribution. Keep in mind, however, that ideal plug flow is a practical impossibility, particularly with highly viscous polymer solutions.

Compare these conclusions with example 7 in chapter 10.

11.5 HETEROGENEOUS STEREOSPECIFIC POLYMERIZATION (4, 5)

The investigators Ziegler and Natta showed that certain combinations of metal alkyls and metal halides would effectively cause the stereospecific polymerization of a variety of monomers by an addition mechanism. A common example of a *Ziegler-Natta catalyst* system is aluminum triethyl and titanium tetrachloride:

$$Al(C_2H_5)_3 + TiCl_4 \rightarrow \text{complex ppt.}$$

When these substances are mixed in an inert solvent, a precipitate is formed along with a highly colored supernatant (deep violet or brown). It is known that the reaction involves the reduction of the titanium to a lower valence state, probably +2, since $TiCl_3$ will also form an effective catalyst. The supernatant liquid alone will polymerize olefins (aliphatic hydrocarbon monomers—e.g., ethylene, propylene), but the resulting polymers show little stereospecificity. The stereo-specific polymerization is thought to occur at active sites on the surface of the precipitate. One postulated mechanism is shown in Figure 11.1, where the growing chain unreels from the active site.

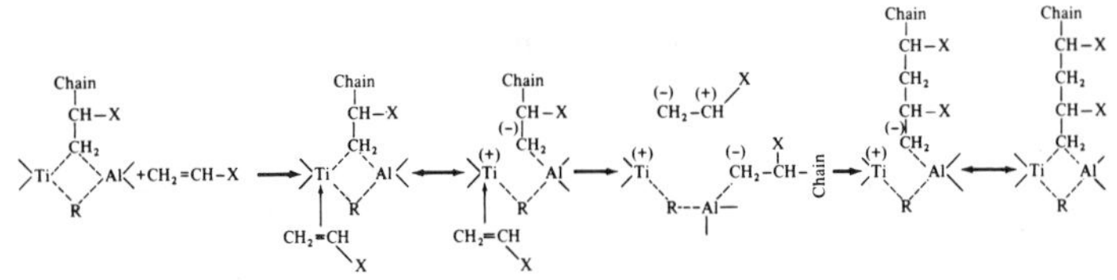

Figure 11.1.   Postulated mechanism for Ziegler-Natta Catalysis (5). (Excerpted by special permission from *Chemical Engineering*, April 2, 1962. Copyright 1962, by McGraw-Hill Book Co., Inc., New York, N.Y. 10036.)

Little is definitely known about the kinetics and mechanisms in Ziegler-Natta catalysis. The results seem to depend on the specific monomer and on the ratio of catalyst components, among other things. Great care must be exercised in the

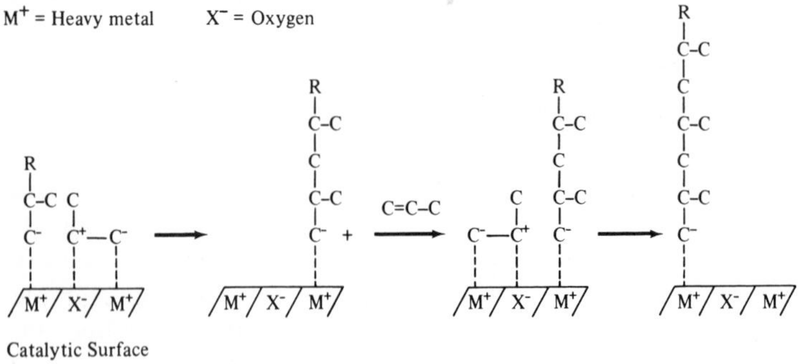

Figure 11.2.   Postulated mechanism for Phillips Catalysis (5). (Excerpted by special permission from *Chemical Engineering*, April 2, 1962. Copyright 1962, by McGraw-Hill Book Co., Inc., New York, N.Y. 10036.)

preparation and use of the catalysts, as they are easily poisoned by oxygen and water and are toxic and highly reactive.

Another type of stereospecific polymerization catalyst was developed by the Phillips Petroleum Company. These Phillips catalysts are CrO_3 supported on silica or alumina. A postulated mechanism for these catalysts is sketched in Figure 11.2.

The catalyst systems described above not only result in the stereospecific polymerization of unsymmetrical monomers but are also capable of producing relatively linear polyethylene (the Phillips catalysts the most linear of all) at low pressures. The older free-radical polymerizations required much higher pressures and gave more highly branched products.

REFERENCES

1. Plesch, P. H., ed. *The Chemistry of Cationic Polymerization.* The Macmillan Company, New York, 1963.

2. Lenz, R. W. *Organic Chemistry of Synthetic High Polymers,* ch. 13. John Wiley & Sons, Inc., New York, 1967.

3. Odian, G. *Principles of Polymerization,* ch. 5. McGraw-Hill Book Co., New York, 1970.

4. Reich, L. and A. Schindler. *Polymerization by Organometallic Compounds.* John Wiley & Sons, Inc., New York, 1966.

5. Guccione, E. Stereospecific catalysts. *Chem. Eng.,* April 2, p. 93, 1962.

12
Copolymerization

12.1 MECHANISM

When two monomers, M_1 and M_2 are copolymerized, there are four possible propagation reactions; $M\cdot$ represents a growing chain radical. (Mechanisms for copolymerization will be illustrated for free-radical addition reactions. The same general sort of thing occurs in ionic systems, with different quantitative parameters.)

reaction	rate equation	
$M_1\cdot + M_1 \xrightarrow{k_{11}} M_1\cdot$	$k_{11}[M_1\cdot][M_1]$	(12-1)
$M_1\cdot + M_2 \xrightarrow{k_{12}} M_2\cdot$	$k_{12}[M_1\cdot][M_2]$	(12-2)
$M_2\cdot + M_2 \xrightarrow{k_{22}} M_2\cdot$	$k_{22}[M_2\cdot][M_2]$	(12-3)
$M_2\cdot + M_1 \xrightarrow{k_{21}} M_1\cdot$	$k_{21}[M_2\cdot][M_1]$	(12-4)

Application of the steady-state assumption to the radicals $M_1\cdot$ and $M_2\cdot$ requires that they must be generated and consumed at equal rates.

$M_1\cdot$'s are generated in reaction (12-4) and consumed in reaction (12-2). Note that (12-1) just converts one $M_1\cdot$ into another one, with no net change in their number. Therefore,

$$k_{12}[M_1\cdot][M_2] = k_{21}[M_2\cdot][M_1]. \qquad (12\text{-}5)$$

The rates of consumption of monomers M_1 and M_2 are:

$$-d[M_1]/dt = k_{11}[M_1\cdot][M_1] + k_{21}[M_2\cdot][M_1] \qquad (12\text{-}6)$$

$$-d[M_2]/dt = k_{12}[M_1\cdot][M_2] + k_{22}[M_2\cdot][M_2] \qquad (12\text{-}7)$$

Dividing (12-6) by (12-7) and eliminating the $[M \cdot]$'s with (12-5) gives

$$\frac{d[M_1]}{d[M_2]} = \frac{[M_1]}{[M_2]} \frac{r_1[M_1] + [M_2]}{[M_1] + r_2[M_1]} \qquad (12\text{-}8)$$

where r_1 and r_2 are the *reactivity ratios.*

$r_1 = k_{11}/k_{12} = $ relative preference of $[M_1 \cdot]$ for M_1/M_2

$r_2 = k_{22}/k_{21} = $ relative preference of $[M_1 \cdot]$ for M_2/M_1

Reactivity ratios are experimentally determined (*1*) or may be estimated (*2*).

This relation may be put in a more convenient form by defining

$f_1 = $ mole fraction of monomer 1 in the reaction mass
 at any instant

$F_1 = $ mole fraction of monomer 1 in the polymer formed
 at that instant.

$$F_1 = 1 - F_2 = \frac{d[M_1]}{d([M_1] + [M_2])} \qquad (12\text{-}9)$$

$$f_1 = 1 - f_2 = \frac{[M_1]}{[M_1] + [M_2]} \qquad (12\text{-}10)$$

Combining (12-8), (12-9) and (12-10) gives

$$F_1 = \frac{r_1 f_1^2 + f_1 f_2}{r_1 f_1^2 + 2 f_1 f_2 + r_2 f_2^2}$$

$$= \frac{(r_1 - 1) f_1^2 + f_1}{(r_1 + r_2 - 2) f_1^2 + 2(1 - r_2) f_1 + r_2} \qquad (12\text{-}11)$$

12.2 SIGNIFICANCE OF REACTIVITY RATIOS

To gain an appreciation for the physical significance of (12-11), let's look at some special cases of the reactivity ratios.

Case 1: $r_1 = r_2 = 0$.

With both reactivity ratios zero, neither radical can regenerate itself; hence, a *perfectly alternating* copolymer results ($F_1 = 0.5$

always) until one of the monomers is used up, at which point polymerization stops.

Case 2: $r_1 = r_2 = \infty$.
Here, $M_1 \cdot$ radicals can only add M_1 monomer, and $M_2 \cdot$ only M_2, so the polymer formed will be a *mixture* of homopolymer 1 and homopolymer 2.

Case 3: $r_1 = r_2 = 1$
Under these conditions, the growing chain radicals can't distinguish between the two monomers, so the addition is *completely random*, depending only on the concentrations of the monomers in the vicinity of the chain ends; i.e., $F_1 = f_1$.

Case 4: $r_1 r_2 = 1$
This is the so-called ideal copolymerization, where each radical displays the same preference for one of the monomers over the other, $k_{11}/k_{12} = k_{21}/k_{22}$. In this case, (12-11) reduces to

$$F_1 = \frac{r_1 f_1}{f_1(r_1 - 1) + 1}. \qquad (12\text{-}12)$$

The reader familiar with distillation theory will note here the exact analogy between (12-12) and the vapor-liquid equilibrium composition relation for ideal solutions with a constant relative volatility.

Case 5: $r_1 < 1$, $r_2 < 1$ (the case where both are greater than 1 does not occur)
This situation corresponds to an azeotrope in vapor-liquid equilibrium. At the azeotropic composition, $F_1 = f_1 = \dfrac{(1 - r_2)}{(2 - r_1 - r_2)}$.

12.3 VARIATION OF COMPOSITION WITH CONVERSION

In general, $F_1 \neq f_1$; i.e., the composition of the polymer formed at any instant will differ from that of the monomer mass. Thus, as the reaction proceeds, the unreacted monomer batch will become depleted in the more reactive monomer, and, as the composition of the unreacted monomer changes, so will that of the polymer being formed, according to (12-11).

Example 1. Draw curves of instantaneous copolymer composition, F_1 versus monomer composition, f_1, for the following systems, and indicate the direction of composition drift as the reaction proceeds.

a. Butadiene (*1*), styrene (*2*), 60C; $r_1 = 1.39, r_2 = 0.78$
b. Vinyl acetate (*1*), styrene (*2*), 60C; $r_1 = 0.01, r_2 = 55$
c. Maleic anhydride (*1*), isopropenyl acetate (*2*), 60C; $r_1 = 0.002, r_2 = 0.032$

Solution. Application of (12-11) gives the plots in Figure 12.1. Note that system (a) approximates ideal copolymer-

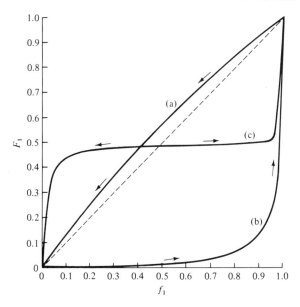

Figure 12.1. Instantaneous copolymer composition (F_1) versus monomer composition (f_1). (a) butadiene (1), styrene (2), 60C; $r_1 = 1.39$, $r_2 = 0.78$. (b) vinyl acetate (1), styrene (2), 60C; $r_1 = 0.01$, $r_2 = 55$. (c) maleic anhydride (1), isopropenyl acetate (2); $r_1 = 0.002$, $r_2 = 0.032$. The direction of composition drift in a batch reactor is indicated by arrows (example 1).

ization (case 4 above). In system (b), styrene is the preferred monomer regardless of the terminal radical; hence, the copolymer is largely styrene until styrene monomer is nearly used up. System (c) approximates case 1 above. The direction of composition drift with conversion is indicated by arrows.

Note that system (c) forms an azeotrope at $F_1 = f_1 = 0.493$.

For $f_{1o} > 0.493$, F_1 and f_1 will increase with conversion, and vice versa.

Consider a batch consisting of a total of M moles of monomer (M = M_1 + M_2). At time t, the monomer has a composition f_1. In the time interval dt, dM moles of monomer polymerize to form dM moles of copolymer with a composition F_1. Therefore, at time $t + dt$, there are left $(M - dM)$ moles monomer whose composition has been changed to $(f_1 - df_1)$. Writing a material balance on monomer 1: M_1 in monomer at t = M_1 in monomer at $(t + dt)$ + M_1 in polymer formed in interval dt

$$f_1 M = (M - dM)(f_1 - df_1) + F_1 dM. \qquad (12\text{-}13)$$

Expanding and neglecting second-order differentials gives

$$\frac{dM}{M} = \frac{df_1}{(F_1 - f_1)}. \qquad (12\text{-}14)$$

At the start of the reaction, there are present M_o moles monomer with a composition f_{1o}, and, at some later time, there are M moles monomer left with a composition f_1. By integrating between these limits,

$$\ln \frac{M}{M_o} = \int_{f_{1o}}^{f_1} \frac{df_1}{(F_1 - f_1)}. \qquad (12\text{-}15)$$

This equation is the exact analog of the Rayleigh equation relating the amount and composition of the still-pot liquid in a batch distillation. By choosing values for f_1 and calculating the corresponding F_1's from (12-11), the integral may be evaluated graphically or numerically, thereby giving a relation between the monomer composition and conversion $\left(1 - \dfrac{M}{M_o}\right)$.

An analytic solution to (12-11 and 12-15) has also been obtained (3). For $r_1 \neq 1$, $r_2 \neq 1$ (if either is equal to 1, see the original reference)

$$\frac{M}{M_o} = \left(\frac{f_1}{f_{1o}}\right)^\alpha \left(\frac{f_2}{f_{2o}}\right)^\beta \left(\frac{f_{1o} - \delta}{f_1 - \delta}\right)^\gamma \qquad (12\text{-}16)$$

where $\alpha = r_2/(1 - r_2)$
$\beta = r_1/(1 - r_1)$

$$\gamma = (1 - r_1 r_2)/(1 - r_1)(1 - r_2)$$
$$\delta = (1 - r_2)/(2 - r_1 - r_2).$$

Knowing the monomer composition as a function of conversion immediately gives the *instantaneous* copolymer composition as a function of conversion through (12-11). This is important to know, because, if there is a large variation in the composition of the copolymer chains formed from the beginning of the reaction to high conversions, there may be a wide variation in their *properties*, also. For example, if there is a large enough variation in the refractive indices of the materials formed at low and high conversions, the resulting copolymer may be hazy even though completely amorphous.

In addition to the instantaneous copolymer composition, F_1, another quantity of interest is $<F_1>$, the *overall average composition* of all the copolymer which has been formed in the batch up to a particular conversion. This is obtained from a material balance:

moles M_1 charged = moles M_1 in polymer + moles M_1 left in monomer

$$f_{1o} M_o = <F_1>(M_0 - M) + f_1 M. \qquad (12\text{-}17)$$

after rearranging,

$$<F_1> = \frac{f_{1o} - f_1 \left(\dfrac{M}{M_o} \right)}{\left(1 - \dfrac{M}{M_o} \right)}. \qquad (12\text{-}18)$$

The distillation analog of this equation tells the well-educated bootlegger how much of his 20-proof sour mash he must distill over to have 100-proof rotgut in the barrel under his condenser.

Example 2. For the styrene-butadiene system of example 1, sketch curves of instantaneous copolymer composition, F_1, and overall average copolymer composition, $<F_1>$ versus conversion, for a batch reaction starting with a 50–50 (mole per cent) initial monomer charge.

Solution. Calculations are facilitated by choosing f_1 as the independent variable. As indicated in example 1, butadiene is the more reactive monomer, so, as the reaction proceeds, both polymer and monomer will be enriched in styrene, so f_1 will vary between 0.5 and 0. For a given value of f_1, the conversion $1 - \dfrac{M}{M_o}$ is calculated from (12-16) with $\alpha = 3.55$,

$\beta = -3.57$, $\gamma = 0.983$ and $\delta = -1.29$. For the chosen value of f_1, F_1 is obtained as in example 1. $\langle F_1 \rangle$ is calculated from (12-18) by using the chosen f_1 and the value of $\dfrac{M}{M_o}$ obtained from (12-16).

The results are plotted in Figure 12.2. Note that, in this par-

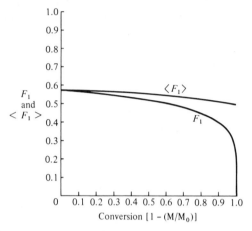

Figure 12.2. Instantaneous (F_1) and overall average ($\langle F_1 \rangle$) copolymer composition versus conversion, $r_1 = 1.39$, $r_2 = 0.78$, $f_{1_o} = 0.50$ (example 2).

ticular case, $\langle F_1 \rangle$ does not vary much between 0 and 100 per cent conversion, but the composition of the chains formed, F_1, changes considerably, particularly at high conversions.

Example 3. Suggest three techniques for producing copolymers of fairly uniform composition—i.e., those in which F_1 does not vary much.

Solution. Three possibilities are sketched in Figure 12.3. With a batch reactor, the more reactive monomer must be replenished as the reaction proceeds to maintain f_1 (and therefore F_1) constant. A method for calculating the appropriate rate of addition has been described (4). In a continuous stirred tank (backmix) reactor, both f_1 and F_1 are constant with time. In a continuous plug-flow reactor, the variation in F_1 can be kept small by limiting the conversion per pass in the reactor. Note that all these techniques require facilities for separating unreacted monomer from the polymer and, in most cases, recycling it.

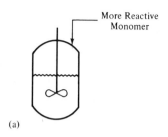

(a)

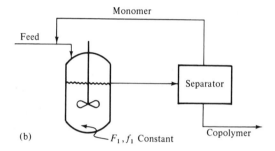

(b)

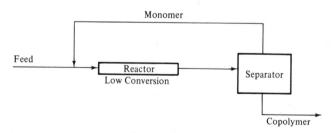

Figure 12.3. Techniques for minimizing the spread of copolymer composition; (a) batch reactor, (b) continuous stirred-tank reactor, (c) tubular reactor.

12.4 COPOLYMERIZATION KINETICS

The problem of copolymerization kinetics is not nearly in such good shape. In addition to the four propagation reactions, there are three possible termination reactions ($M_1 \cdot +$

$M_2\cdot$, $M_1\cdot + M_1\cdot$, $M_2\cdot + M_2\cdot$), each with its own rate constant. A general rate equation has been developed (5), but, because of a lack of independent knowledge of the constants involved and the mathematical complexity, it hasn't been used much. An approximate integrated expression including first-order initiator decay has been presented recently (6). It involves a single, average termination rate constant. This constant would not be expected to remain constant as composition varies. Nevertheless, there appears to be some experimental verification of the relation. In certain instances where the compositions aren't varying too much (e.g., in the control of a continuous backmix reactor), simplification of (12-6) and (12-7) to

$$-d[M_1]/dt = K_1[M_1] \qquad (12\text{-}19)$$

$$-d[M_2]/dt = K_2[M_2] \qquad (12\text{-}20)$$

where the constants $K_1 = (k_{11}[M_1\cdot] + k_{21}[M_2\cdot])$ and $K_2 = (k_{12}[M_1\cdot] + k_{22}[M_2\cdot])$ are determined experimentally may prove satisfactory.

REFERENCES

1. Mark, H. et. al. Copolymerization Reactivity Ratios, ch. II-6 in *Polymer Handbook* (J. Brandrup and E. H. Immergut, eds.). John Wiley & Sons, Inc., New York, 1966.

2. Alfrey, T., Jr. and C. C. Price. Relative reactivities in vinyl copolymerization. *J. Polymer Sci.*, **2**, 101, 1947.

3. Meyer, V. E. and G. G. Lowry, Integral and differential binary copolymerization equations. *J. Polymer Sci.*, **A3**, p. 2843, 1965.

4. Hanna, R. J. Synthesis of chemically uniform copolymers. *Ind. and Eng. Chem.*, **49**, No. 208, 1957.

5. DeButts, E. H. Integration of copolymerization rate equations. *J. Amer. Chem. Soc.*, **72**, No. 411, 1950.

6. O'Driscoll, K. F. and R. S. Knorr. Multicomponent Polymerization I. Integration of the Rate Equations. *Macromol.* **1**, No. 367, 1968.

13
Polymerization Practice

13.1 BULK POLYMERIZATION

The simplest and most direct method of converting monomer to polymer is known as *bulk* or *mass* polymerization. A typical charge might consist of monomer, a monomer-soluble initiator and perhaps a chain-transfer agent.

As simple as this sounds, there are some serious difficulties which can be encountered. One of them is illustrated in Figure 13.1 (*1*) which indicates the course of polymerization for various concentrations of methyl methacrylate (Lucite, Plexiglas, Perspex) in benzene, an inert solvent. The reactions were carefully maintained at constant temperature. At the lower

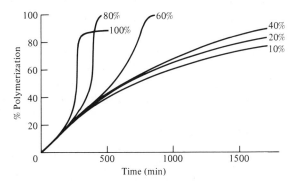

Figure 13.1. Polymerization of methyl methacrylate at 50C in the presence of benzoyl peroxide at various concentrations of monomer in benzene. (Reprinted from Paul J. Flory: Principles of Polymer Chemistry. Copyright 1953, by Cornell University. Used by permission of Cornell University Press.)

monomer concentrations, the conversion versus time curves are described well by (10-14). At the higher concentrations, however, a distinct acceleration of the rate of polymerization is observed, which does not conform to the classical kinetic scheme. This phenomenon is known variously as *autoacceleration*, the *gel effect*, or the *Tromsdorff effect*.

The reasons for this behavior lie in the difference between the propagation reaction (10-3) and the termination steps (10-4a,b), and the extremely high viscosities of concentrated polymer solutions (10^6 poise might be a 'ballpark' figure). The propagation reaction involves the approach of a small monomer molecule and a growing chain end, whereas termination requires that the ends of two growing chains get together. At high polymer concentrations, it becomes exceedingly difficult for the growing chain ends to drag their chains through the entangled mass of dead polymer chains. It is nowhere near as difficult for a monomer molecule to pass through the reaction mass. Thus, the rate of the termination reaction becomes limited not by the nature of the chemical reaction but by the rate at which the reactants can diffuse together to react—i.e., *diffusion controlled*. This lowers the effective rate constant k_t, and, since it appears in the denominator of (10-11), the net effect is to increase the rate of polymerization. At very high polymer concentrations and below the temperature at which the chains become essentially immobile (T_g of the monomer-plasticized polymer) even the propagation reaction is diffusion limited—hence, the leveling off of the 100 per cent curve.

The difficulties are compounded by the inherent nature of the reaction mass. Vinyl monomers have rather large exothermic heats of polymerization, typically between -10 and -21 Kcal/gmole. Organic systems also have low heat capacities and thermal conductivities, about half those of aqueous systems. To top it all off, the tremendous viscosities prevent effective convective (mixing) heat transfer. As a result, overall heat-transfer coefficients on the order of 1 Btu/hr, ft^2, °F are not uncommon, making it difficult to remove the heat generated by the reaction. This raises the temperature, further increasing the rate of reaction and heat evolution, etc., and can ultimately lead to disaster. To quote Schildknecht on laboratory bulk polymerizations (2), "If a complete rapid polymerization of a reactive monomer in large bulk is attempted, it may lead to loss of the apparatus, the polymer or even the experimenter."

Example 1. The *maximum possible* temperature rise in a polymerizing batch may be calculated by assuming that no heat is transferred from the system—i.e., the adiabatic temperature rise. Estimate the adiabatic temperature rise for the bulk polymerization of styrene, $\Delta H_p = -16.4$ Kcal/gmole, molecular weight = 104.

Solution. The polymerization of one g mole of styrene liberates 16,400 cal (assuming complete conversion). In the absence of heat transfer, all this energy heats up the reaction mass. The heat capacities of organic compounds are often difficult to find, and, since the reaction mass is going from monomer to polymer, which in general have different heat capacities, the heat capacity of the reaction mass changes with conversion and probably also with temperature. To a good approximation, however, the heat capacity of most liquid organic systems may be taken as 0.5 cal/*gram*, °C. Thus,

$$\Delta T_{max} = \left(16,400\,\frac{cal}{g\,mole}\right)\left(\frac{1\,g\,mole}{104\,g}\right)\left(\frac{g\,°C}{0.5\,cal}\right) \simeq 315°C(!)$$

Keep in mind that the normal boiling point of styrene is 146C.

These problems are circumvented in a number of ways:

a. *by keeping at least one dimension of the reaction mass small, permitting heat to be conducted out.* Poly methylmethacrylate sheets are cast between glass plates at a maximum thickness of 3/4 inch.

b. *by maintaining very low reaction rates through low temperatures and initiator concentrations.* The polymerization times for the sheets in (a) are on the order of 30–100 hours, and the temperatures are raised slowly as the monomer concentration drops. This approach has obvious economic disadvantages.

c. *by starting with a* sirup *instead of the pure monomer.* A sirup is simply a solution of the polymer in the monomer. It can be made in either of two ways: (1) by carrying the monomer to partial conversion in a kettle, or (2) by dissolving preformed polymer in monomer. Starting off with a sirup means that some of the conversion has already been accomplished, cutting heat generation and monomer concentration in the final polymerization. Since the density of a reaction mass increases on the order of 20–30 percent between 0 and 100 percent conversion in a polymerization, the use of a sirup has the added advantage of cutting shrinkage when casting a polymer.

d. *by carrying out the reaction continuously, with a lot of heat-transfer surface per unit conversion.*

Batch bulk polymerization is generally used to obtain objects with desired shape by polymerizing in a mold. Examples are casting, potting and encapsulation of electrical components and impregnation of reinforcing agents followed by polymerization. It is also used extensively for the production of thermosetting resins, which are carried to a conversion short of complete crosslinking in the reactor. The crosslinking is later completed in a mold.

Continuous bulk polymerization is becoming increasingly important in the production of thermoplastic molding compounds. A continuous bulk process is outlined in Figure 13.2 (*3*). Conversion is carried to about 40 percent in a stirred tank. The reaction mass then passes down a tower with the temperature increasing to keep the viscosity to a manageable level and to drive up the conversion. The tower may be a

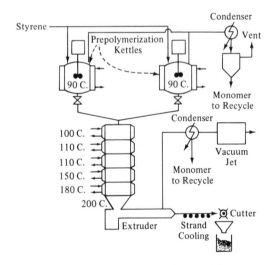

Figure 13.2.   Continuous bulk polymerization of styrene. (Reprinted by special permission from *Chemical Engineering,* August 1, 1962. Copyright 1962, by McGraw-Hill Book Co., Inc., New York, N.Y. 10036.)

simple gravity-flow device, or it may contain slowly rotating spiral blades which scrape the walls, promoting heat transfer and conveying the reaction mass downward. The reaction mass is fed from the tower to a vented extruder at better than 95 percent conversion. Some additional conversion takes place in the extruder, and a vacuum sucks off unreacted monomer, which may be recycled. The extruded strands of molten polymer are then water cooled and chopped to form the roughly $1/8 \times 1/8 \times 1/8$ inch pellets which are sold to processors as molding powder. There has also been a great deal of recent interest in continuous sheet-casting processes.

The advantages of bulk polymerization follow.

1. As only monomer, initiator and perhaps chain-transfer agent are used, the purest possible polymer is obtained. This can be important in electrical and optical applications.

2. Objects may be conveniently cast to shape. If the polymer is one which is crosslinked in the synthesis reaction, this is the *only* way of obtaining such objects short of machining from larger blocks.

3. Bulk polymerization provides the greatest possible polymer yield per reactor volume.

Some of its disadvantages are the following.

1. It is often difficult to control.

2. To keep it under control, it may have to be run slowly, with the attendant economic disadvantages.

3. As indicated by (10-11) and (10-21), it may be difficult to get both high rates and high average chain lengths because of the opposing effects of [I].

4. It can be difficult to remove the last traces of unreacted monomer. This can be extremely important—for example, if the polymer is intended for use as a food wrap and FDA clearance is required.

13.2 SOLUTION POLYMERIZATION

The addition of an inert solvent to a bulk polymerization mass minimizes many of the difficulties encountered in bulk systems. As shown in Figure 13.1, it reduces the tendency toward autoacceleration. The inert diluent adds its heat capacity without contributing to the evolution of heat, and it cuts the viscosity of the reaction mass at any given conversion. In addition, the heat of polymerization may be conveniently and efficiently removed by refluxing the solvent. Thus, the danger of runaway reactions is minimized.

Example 2. Estimate the adiabatic temperature rise for the polymerization of a 20 percent (by weight) solution of styrene in an inert organic solvent.

Solution. In 100 g of the reaction mass, there are 20 g of styrene, so the energy liberated on its complete conversion to polymer is

$$(20 \text{ grams})\left(\frac{1 \text{ g mole}}{104 \text{ g}}\right)\left(\frac{16,400 \text{ cal}}{\text{g mole}}\right) = 3,150 \text{ cal}.$$

The adiabatic temperature rise is then

$$\Delta T_{max} = (3,150 \text{ cal})\left(\frac{\text{g} \,^\circ\text{C}}{0.5 \text{ cal}}\right)\left(\frac{1}{100 \text{ g}}\right) = 63 \text{C}.$$

The advantages of solution polymerization follow.

1. Heat removal and control are easier.

2. Since the reactions are more likely to follow known theoretical kinetic relations, the design of reactor systems is facilitated.

3. For some applications (e.g., lacquers), the desired polymer solution is obtained directly from the reactor.

Some of its disadvantages are the following.

1. Since both rate and average chain length are proportional to [M] (in free-radical addition), the use of a solvent lowers them. Additional lowering of $\bar{x}_n$ will occur if the solvent acts as a chain-transfer agent.

2. Large amounts of expensive, flammable and perhaps toxic solvent are required.

3. Separation of the polymer and recovery of the solvent require additional technology.

4. Removal of the last traces of the solvent may be difficult (see 4 under bulk).

5. Use of an inert in the reaction mass lowers the yield per volume of reactor.

Ionic and heterogeneous stereospecific polymerizations are almost exclusively solution processes (*4, 5, 6, 7*). The cationic polymerization of isobutylene with BF_3 is carried out at -150F

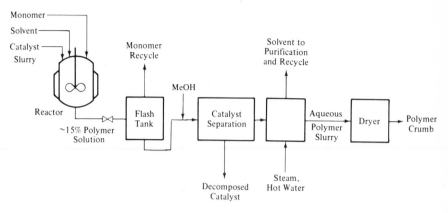

Figure 13.3. Typical Ziegler-Natta polymerization process.

with ethylene as a refluxing solvent. The temperature is controlled by manipulating the pressure on the boiling system. Figure 10.3 sketches a typical process utilizing a Zeigler-Natta catalyst system. Heat removal from the reactor(s) may be accomplished by refluxing the solvent, cooling jackets or external pump-around heat exchangers, or combinations of these. Where a crystalline polymer is being produced, the reaction may be carried out either above or below the crystalline melting point of the polymer. In the latter case, the polymer will precipitate out as it is formed, and the product of the reactor is a slurry rather than a homogeneous solution. The catalyst is normally deactivated with methanol and filtered, centrifuged or settled out. Solvent and unreacted monomer are stripped with hot water and steam and recovered, leaving a water slurry of polymer which is then dried to form a crumb. With rubbers, the crumb is compacted and baled; with plastics, it is normally extruded and pelletized. Reactor designs for these processes are interesting and extremely varied. A majority of the newer processes are continuous.

13.3 SUSPENSION, BEAD OR PEARL POLYMERIZATION

When discussing bulk polymerization, it was mentioned that one of the ways of facilitating heat removal was to keep one dimension of the reaction mass small. This is carried to its logical extreme in suspension polymerization by suspending the monomer in the form of droplets 0.01–1 mm in diameter in an inert, nonsolvent liquid (almost always water). In this way, each droplet becomes a single bulk reactor, but with dimensions small enough so that heat removal is no problem and the heat can easily be soaked up by and removed from the low-viscosity, inert suspension medium.

An important characteristic of these systems is that the suspensions are *thermodynamically unstable* and must be maintained with agitation and suspending agents. A typical charge might consist of:

monomer (water insoluble)
initiator (monomer soluble) } monomer phase
chain-transfer agent (monomer soluble)

water

suspending agent - { protective colloid
                    insoluble inorganic salt

Two types of suspending agent are used. A protective colloid is a water-soluble polymer whose function is to increase the viscosity of the continuous water phase. This hydrodynamically hinders coalescence of monomer drops but is inert with regard to the polymerization. A finely divided insoluble inorganic salt such as $MgCO_3$ may also be used. It collects at the droplet-water interface by surface tension and prevents coalescence of the drops upon collision.

The monomer phase is suspended in the water at about a $1/2$ to $1/4$ monomer/water volume ratio. The reactor is purged with nitrogen and heated to start the reaction. Once underway, temperature control in the reactor is facilitated by the added heat capacity of the water, the low viscosity of the reaction mass—essentially that of the continuous phase—allowing heat removal through a jacket and the possibility of refluxing the monomer and water.

The size of the product beads depends on the strength of agitation. Between about 20 and 70 percent conversion, *agitation is extremely critical*. Below 20 percent, the organic phase is still fluid enough to redisperse; above 70 percent, the particles are rigid enough to prevent agglomeration, but, if agitation stops or weakens between these limits, the sticky particles will coalesce or agglomerate in a large mass and finish polymerization that way. Schildknecht (8) observed that "After such uncontrollable polymerization is completed in an enormous lump, it may be necessary to resort to a compressed air drill or other mining tools to salvage the polymerization equipment."

Since any flow system is bound to have some relatively stagnant corners, it has been impractical to run suspension polymerization continuously. Figure 13-4 (9) shows a typical

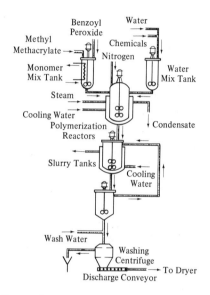

Figure 13.4. Suspension polymerization of methyl methacrylate. (Excerpted by special permission from *Chemical Engineering*, June 6, 1966. Copyright 1966, by McGraw-Hill Book Co., Inc., New York, N.Y. 10036.)

process. The reactors are usually jacketed, glass-lined steel kettles of up to 20,000 gal capacity. The polymer beads are filtered or centrifuged and either water washed to remove the protective colloid or rinsed with a dilute acid to decompose the $MgCO_3$. The beads are quite easy to handle when wet, but they tend to pick up a static charge when dry, making them cling to each other and everything else. The beads can be molded directly, extruded and chopped to form molding powder or occasionally used as is—e.g., ion-exchange resins.

The major advantages of suspension polymerization, then, are the following.

1. Heat removal and control are easy.

2. The polymer is obtained in a convenient, easily handled form.

The disadvantages include the following.

1. There is a low yield per reactor volume.

2. A somewhat less pure polymer than from bulk polymerization results, since there are bound to be remnants of the suspending agent(s) absorbed on the particle surface.

3. The process cannot be run continuously, although, if several reactors are alternated, the process may be continuous from that point on.

13.4 EMULSION POLYMERIZATION

When the supply of natural rubber from the East was cut off by the Japanese in World War II, the United States was left without this essential material. The success of the Rubber Reserve Program in developing a suitable synthetic substitute and the facilities to produce it in the necessary quantities is one of the all-time outstanding accomplishments of chemists and engineers. The styrene-butadiene copolymer rubber GR-S (Government Rubber-Styrene) or SBR (Styrene-Butadiene Rubber) as it is now called (developed during the war) is still the most important synthetic rubber and is still produced, along with a variety of other polymers, largely by the emulsion polymerization process developed then.

A rational explanation of the mechanism of emulsion polymerization was developed by Smith and Ewart after the war, when the information gathered at various locations could be freely examined and exchanged (*10*). Perhaps the best way of introducing the subject is to list a typical reactor charge:

Typical Emulsion Polymerization Charge
100 parts (by weight) monomer (water insoluble)
180 parts water
2–5   fatty acid soap
0.1–0.5   *water-soluble* initiator
0–1   chain-transfer agent (monomer soluble)

The first question that comes up is "what's the soap for?" Soaps are the sodium or potassium salts of organic acids or sulfates:

$$[R—\overset{\overset{\textstyle O}{\|}}{C}—O]^- \, Na^+.$$

When they are added to water in low concentrations, they ionize and float around freely much as sodium chloride ions would. The anions, however, consist of a highly polar hydrophilic (water-seeking) head and an organic, hydrophobic (water-fearing) tail. As the soap concentration is increased, a value is suddenly reached where the anions begin to agglomerate in *micelles* rather than float around individually. These micelles have dimensions on the order of 50–60Å (1 cm = $10^4\mu = 10^8$ Å), far too small to be seen with a light microscope. They consist of a tangle of the hydrophobic tails in the interior (getting as far away from the water as possible) with the hydrophilic heads on the outside. This process is easily observed by following the variation of a number of solution properties with soap concentration—e.g., electrical conductivity or surface tension (Figure 13.5). The break occurs when micelles

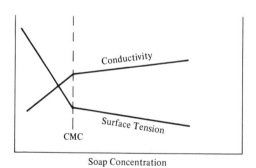

Figure 13.5. Variation of solution properties at the critical micelle concentration.

start to form and is known as the *critical micelle concentration* or CMC.

When an organic monomer is added to an aqueous micelle solution, it naturally prefers the organic environment at the interior of the micelles. Some of it congregates there, swelling the micelles until an equilibrium is reached with the contraction force of surface tension. *Most* of the monomer, however, is distributed in the form of *much* larger (1μ or 10^4 Å) droplets stabilized by soap. This complex mixture is an *emulsion*. In contrast to a suspension, a proper emulsion is *thermodynamically stable* and will remain so even if the agitation is stopped. The cleaning action of soaps depends on their ability to emulsify oils and greases.

Despite the fact that most of the monomer is present in the droplets, the swollen micelles, because of their much smaller size, present a much larger surface area than the droplets. This is easily seen by assuming a micelle volume to drop volume ratio of $1/10$ and using the ballpark figures given above. Since the surface/volume ratio of a sphere is $3/R$,

$$\frac{S_{\text{micelle}}}{S_{\text{droplet}}} = \left(\frac{V_{\text{mic.}}}{V_{\text{drop.}}}\right)\left(\frac{R_{\text{drop.}}}{R_{\text{mic.}}}\right) \simeq \left(\frac{1}{10}\right)\left(\frac{10^4}{0.5 \times 10^2}\right) \simeq 20.$$

Free radicals are generated in the aqueous phase by the decomposition of water-soluble initiators, usually potassium or ammonium persulfate:

$$S_2O_8^= \rightarrow 2SO_4^{\overline{\cdot}}$$

persulfate   sulfate ion-radical

Redox systems, so called because they involve the alternate oxidation and reduction of a trace catalyst, are a newer and more efficient means of generating radicals:

1. $S_2O_8^= \quad + \; HSO_3^- \rightarrow \; SO_4^= \; + \; SO_4^{\overline{\cdot}} \; + \; HSO_3\cdot$

    persulfate   bisulfite

2. $S_2O_8^= \quad + \; Fe^{++} \quad\;\; \rightarrow \; SO_4^= \; + \; Fe^{3+} \; + \; SO_4^{\overline{\cdot}}$

3. $HSO_3^- \quad + \; Fe^{3+} \quad\;\; \rightarrow \; HSO_3\cdot \; + \; Fe^{++}$

$S_2O_8^= \quad\;\;\; + \; HSO_3^- \rightarrow \; SO_4^= \; + \; SO_4^{\overline{\cdot}} + HSO_3\cdot$

The original wartime GR-S polymerization was carried out at 50C with potassium persulfate initiator ('hot' rubber). The use of the more efficient redox systems allowed a reduction in polymerization temperature to 5C ('cold' rubber). The latter product has superior properties because the lower polymerization temperature promotes cis 1-4 addition of the butadiene.

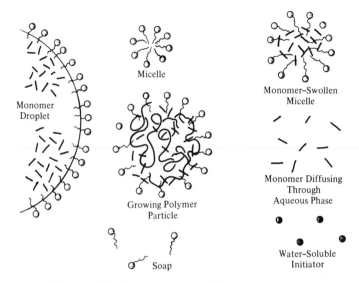

Monomer Droplet

Micelle

Monomer-Swollen Micelle

Monomer Diffusing Through Aqueous Phase

Growing Polymer Particle

Soap

Water-Soluble Initiator

Figure 13.6. Structures in emulsion polymerization.

Figure 13.6 illustrates the structures present during emulsion polymerization.

The radicals thus generated in the aqueous phase bounce around until they encounter some monomer. Since the surface area presented by the monomer-swollen micelles is so much greater than that of the droplets, the probability of a radical entering a monomer-swollen micelle rather than a droplet is large. As soon as the radical encounters the monomer within the micelle, it initiates polymerization. The conversion of monomer to polymer within the growing micelle lowers the monomer concentration therein, and monomer begins to diffuse from uninitiated micelles and monomer droplets to the growing, polymer-containing micelles. Those monomer-swollen micelles not struck by a radical during the initial stages of conversion thus disappear, losing their monomer to those that have been initiated and stabilizing the number of growing particles in the system. The reaction mass now consists of a stable number of growing polymer particles (originally micelles) and the monomer droplets, which simply act as reservoirs supplying monomer to the growing polymer particles by diffusion through the water. The monomer concentration in the growing particles reaches a dynamic equilibrium value dictated by the tendency toward further dilution (increasing the entropy) and the opposing effect of surface tension attempting to minimize the surface area. (Although organic monomers are normally thought of as being water insoluble, their concentrations in the aqueous phase, though small, are sufficient to permit a high enough diffusion flux to maintain the monomer concentration in the polymerizing particles (*11*).)

A monomer-swollen micelle which has been struck by a radical now contains *one* growing chain. With only one radical per particle, there is no way in which the chain can terminate, and it continues to grow until a second radical enters the particle. It may be shown that, under the conditions prevailing within the particle, the rate of termination is much greater than the rate of propagation, so the chain growth is terminated essentially immediately after the entrance of the second radical (*11*). The particle then remains dormant until a third radical enters, initiating the growth of a second chain. This second chain grows until it is terminated by the entry of the fourth radical, and so on.

13.5 KINETICS OF EMULSION POLYMERIZATION

Thus, according to classical emulsion polymerization theory, beyond the first few percent conversion, the reaction mass consists of a stable number of monomer-swollen polymer particles which are the loci of all polymerization. At any given time, a particle contains either one growing chain or no growing chains. Statistically, then, if there are N particles per liter of reaction mass, there are $N/2$ growing chains per liter of reaction mass.

The polymerization rate is given by

$$r_p = k_p [M][M \cdot] \qquad (10\text{-}6)$$

where k_p is the *homogeneous* propagation rate constant (liter of particles/mole sec) and M is the equilibrium monomer concentration *within a particle* (moles/liter of particles). Now, $[M \cdot] = N/2$ moles/liter of *reaction mass* (particles plus water) and the rate of polymerization is

$$r_p = k_p\left(\frac{N}{2}\right)[M] \tag{13-1}$$

$$\left(\frac{\text{moles}}{\text{liter of reaction mass, sec}}\right)$$

$$= \left(\frac{\text{liter of particles}}{\text{mole, sec}}\right)\left(\frac{\text{moles}}{\text{liter of reaction mass}}\right)\left(\frac{\text{moles}}{\text{liter of particles}}\right).$$

Equation (13-1) thus gives the rate of polymerization per total volume of reaction mass. Surprisingly, it predicts the rate to be independent of initiator concentration. Moreover, since N becomes constant beyond the first few percent conversion and [M] is constant, a constant rate is predicted. This is borne out experimentally as shown in Figure 13.7 (12). Deviations from

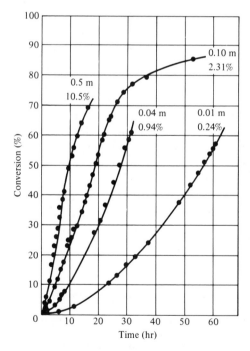

Figure 13.7.   The emulsion polymerization of isoprene as a function soap (potassium laurate) concentration (12). (Copyright 1947, by the American Chemical Society. Reprinted by permission of the copyright owner.)

linearity are observed at low conversions as N is being stabilized and at high conversions when the monomer droplets are
used up and are no longer able to supply the monomer necessary to maintain [M] constant in the growing particles. *Note
the increase in rate with soap concentration.* The more soap
used, the more micelles are established initially, and the higher
N will be. The predicted independence of rate on initiator
concentration must be viewed with caution. It's O.K. as long
as N is held constant, but, in practice, N increases with $[I_o]$.
The more initiator added at the start of a reaction, the greater
the number of monomer-swollen micelles which start growing
before N is stabilized. Methods are available for estimating
N *(10, 11)*.

Example 3. The Putrid Paint Division of Crud Chemicals,
Inc., has available a 10 percent (by weight) of polyvinyl acetate
latex containing 1×10^{14} particles/cc. To obtain optimum
characteristics as an interior wall paint, a larger particle size
and higher concentration of polymer are needed. It is proposed to obtain these by adding an additional 4 parts (by
weight) of monomer per part of polymer to the latex and
polymerizing without further addition of soap. The reaction
will be carried to 85 percent conversion, and the unreacted
monomer will be steam-stripped and recovered. (The conditions of this *seeded* polymerization are set up so that the
classical emulsion polymerization theory applies.)

Estimate the time required for the reaction and the rate of
heat removal (in Btu/gal of original latex, per hour) necessary
to maintain an essentially isothermal reaction at 60C. At what
conversion would the monomer droplets disappear and the
rate cease to be linear?

$$\text{Data:} \quad k_p = 3,700 \frac{\text{liter}}{\text{mole, sec}} \text{ at 60C}$$

$$\Delta H_p = -21 \text{ Kcal/mole monomer unit}$$

$$\rho \text{ (polymer)} = 1.2 \text{ g/cc}, \rho \text{ (monomer)} = 0.8 \text{ g/cc}$$

Concentration of monomer in growing polymer particles =
10 percent (by weight).

Solution. It must first be assumed that, in the absence of
additional soap, no new micelles will be established, so the

original latex particles act as the exclusive loci for further polymerization. Then, [M] and N must be obtained for use in (13-1).

Basis: 100 g monomer-swollen particles − 90 g polymer, 10 g monomer.

$$(90\text{ g})\left(\frac{1\text{ cc}}{1.2\text{ g}}\right) = 75\text{ cc polymer}$$

$$(10\text{ g})\left(\frac{1\text{ cc}}{0.8\text{ g}}\right) = \frac{12.5\text{ cc monomer}}{87.5\text{ cc}}$$

assumes additivity of volume

Molecular weight of vinyl acetate $\left(\begin{array}{c}\text{H} \quad \text{H}\\ | \quad\ | \\ \text{C}\!=\!\text{C} \\ | \quad\ | \\ \text{H} \quad \text{O}\end{array}\right) = 86\ \dfrac{\text{g}}{\text{g mole}}$

$$\begin{array}{c}| \\ \text{C}\!=\!\text{O} \\ | \\ \text{CH}_3\end{array}$$

$$(10\text{ g})\left(\frac{1\text{ g mole}}{86\text{ g}}\right) = 0.116\text{ g moles monomer present}$$

$$[M] = (0.116\text{ g moles monomer})\left(\frac{1}{87.5\text{ cc of particle}}\right)\left(\frac{1000\text{ cc}}{\text{liter}}\right)$$

$$= 1.33\ \frac{\text{g moles monomer}}{\text{liter of particles}}$$

$$N = 1 \times 10^{17}\ \frac{\text{particles}}{\text{liter of original latex}}$$

$$\frac{N}{2} = \frac{1 \times 10^{17}}{2}\ \frac{\text{free radicals}}{\text{liter of original latex}}$$

Now, 1 g mole of free radicals = 6×10^{23} free radicals (Avogadro's no.).

$$\frac{N}{2} = \left(\frac{1 \times 10^{17}}{2}\ \frac{\text{free rads}}{\text{liter orig. latex}}\right)\left(\frac{1\text{ g mole free rads}}{6 \times 10^{23}\text{ free rads}}\right)$$

$$= \frac{1}{12} \times 10^{-6}\ \frac{\text{g moles free rads}}{\text{liter orig. latex}}$$

$$r_p = k_p\frac{N}{2}[M]$$

$$= \left(3700 \, \frac{\text{liter particles}}{\text{g mole, sec}}\right)\left(\frac{1}{12} \times 10^{-6} \, \frac{\text{g moles free rads}}{\text{liter orig. latex}}\right)$$

$$\cdot \left(1.33 \, \frac{\text{g moles monomer}}{\text{liter particles}}\right)$$

$$= 4.1 \times 10^{-4} \, \frac{\text{g moles monomer}}{\text{liter orig. latex, sec}} = 1.48 \, \frac{\text{g moles monomer}}{\text{liter orig. latex, hr}}.$$

By again assuming additivity of volumes, the density of the original latex is found to be 1.02 g/cc.

In *one liter* of the original latex = 1020 grams, there are 102 grams polymer, to which are added $102 \times 4 = 408$ grams monomer, or

$$(408 \text{ grams})\left(\frac{1 \text{ g mole}}{86 \text{ grams}}\right) = 4.74 \text{ g moles monomer}.$$

The reaction converts $4.74 \times 0.85 = 4.02$ g moles monomer which takes $\left(\dfrac{4.02 \text{ g moles monomer}}{\text{liter orig. latex}}\right)\left(\dfrac{1 \text{ liter orig. latex, hr}}{1.48 \text{ g moles monomer}}\right)$

$= 2.72$ hr.

For an isothermal reaction, the rate of heat generation = rate of heat removed

$$= \left(21,000 \, \frac{\text{cal}}{\text{g mole monomer}}\right)\left(1.48 \, \frac{\text{g moles monomer}}{\text{liter orig. latex, hr}}\right)$$

$$= 31,000 \, \frac{\text{cal}}{\text{liter orig. latex, hr}} = 467 \, \frac{\text{Btu}}{\text{hr, gal orig. latex}}.$$

When the monomer droplets just disappear, all the monomer and polymer will be in the swollen polymer particles, a total of 102 g (original polymer) + 408 g (added monomer) = 510 g per liter of original latex. At this point, the particles still contain 10 percent monomer, so there are $(0.1)(510) = 51$ g unconverted monomer. Therefore,

$$\text{conversion} = \left(1 - \frac{M}{M_o}\right) = 1 - \frac{51}{408} = 0.875 \text{ or } 87.5 \text{ percent}.$$

Beyond this point, the rate will drop off as the monomer concentration in the particles falls.

Despite the secondary effect of $[I_o]$ on the rate, it has a strong influence on the average chain length. The greater the rate of radical generation, the greater will be the frequency of alternation between growth and death in a particle, resulting in a lower chain length. If r_c represents the rate of radical capture/liter of reaction mass (half of which produce dead chains), in the absence of transfer

$$\bar{x}_n = \frac{k_p \left(\dfrac{N}{2}\right) [M]}{r_c/2} = k_p \frac{N}{r_c} [M]. \qquad (13\text{-}2)$$

Although the classical theory of emulsion polymerization described above provides a rational and often accurate description of the process, more recent work has indicated that there might be numerous exceptions to it. For example, larger particles may contain two or more growing chains for appreciable lengths of time, some homogeneous initiation may occur in the aqueous phase, and the actual polymerization may take place near the outer surface of a growing particle. Additional work is concerned with elucidating the initial stage of the reaction, which is not treated in detail by the classical theory.

A typical emulsion process is shown in Fig. 13.8. The re-

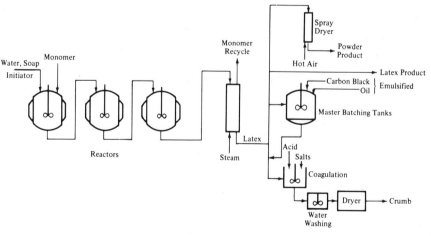

Figure 13.8. Emulsion polymerization process.

actors are usually glass-lined steel, similar to those used for suspension polymerization. Since the emulsion is perfectly stable, emulsion polymerization can be run continuously, and the new processes often have several continuous-stirred-tank reactors in series.

13.6 LATEX TECHNOLOGY

The product of an emulsion polymerization is a *latex*-polymer particles on the order of 500–1500A (0.05 to 0.15μ) stabilized by the soap. A rough calculation shows that even the smallest particle contains many polymer chains.

Example 4. Estimate the number of polymer chains in a latex particle 0.05μ in diameter. Assume particle density = 1 g/cc, average chain length $\bar{x}_n$ = 1000, molecular weight of monomer = 100.

Solution. Volume of particle = $\frac{4}{3} \pi (2.5 \times 10^{-6} \text{ cm})^3 \cong$ 60 $\times$ 10^{-18} cm^3 with a density of 1, each particle weighs 60 $\times$ 10^{-18} grams, and contains (60 $\times$ 10^{-18} grams) $\left(\dfrac{1 \text{ g mole}}{100 \text{ g}}\right)$ = 0.6 $\times$ 10^{-18} g moles monomer units.

There are

$$\left(0.6 \times 10^{-18} \frac{\text{g moles monomer units}}{\text{particle}}\right) \times$$

$$\left(6 \times 10^{23} \frac{\text{monomer units}}{\text{g mole monomer units}}\right) \times$$

= 3.6 $\times$ 10^5 monomer units/particle;

since $\bar{x}_n$ = 1000 $\dfrac{\text{monomer units}}{\text{chain}}$,

there are $\left(3.6 \times 10^5 \dfrac{\text{monomer units}}{\text{particle}}\right)\left(\dfrac{1 \text{ chain}}{1000 \text{ monomer units}}\right) =$

360 $\dfrac{\text{chains}}{\text{particle}}$.

These latices are often important items of commerce in their own right. Two familiar examples are white glue and the so-called water-soluble paints. The latter, also containing pigments and various agents to control application characteristics, are only water soluble (really water dispersible) as long as the latex is maintained. As soon as the water is evaporated and the polymer particles coalesce, one can no longer clean his brushes with water. Where the polymer must be mixed with other materials, the process of *master batching* sometimes allows this to be done conveniently and uniformly. In rubber technology, carbon black and oil are emulsified and mixed with the rubber latex, and then the whole works is coagulated together, giving a uniform and intimate dispersion of the additives in the rubber.

For many applications, the solid polymer must be recovered from the latex. The simplest method is spray drying, but, since no attempt is made to remove the soap, the product is an extremely impure polymer. A latex may be *creamed* by adding a material which is at least a partial solvent for the polymer— e.g., acetone. This makes the particles sticky and causes some agglomeration. The latex is then coagulated by adding an acid (e.g., sulfuric), which converts the soap to the insoluble hydrogen form, or by adding an electrolyte salt, which builds up charges on the particles, causing them to agglomerate through electrostatic attraction. The former method leaves much insoluble soap adsorbed on the particle surfaces. The coagulated polymer *crumb* is then washed, dried and either baled or processed further.

In summary, the advantages of emulsion polymerization are the following.

1. It is easy to control. The viscosity of the reaction mass is much less than a true solution of comparable concentration, the water adds its heat capacity, and the reaction mass may be refluxed.

2. It is possible to obtain both high rates of polymerization and high average chain lengths through the use of high soap and low initiator concentrations.

3. The latex product is often directly valuable or aids in obtaining uniform compounds through master batching.

Its disadvantages are the following.

1. It is difficult to get pure polymer. The tremendous surface area of the tiny particles provides plenty of room for adsorbed impurities—this includes water attracted by residual soap, traces of which can cause problems in certain applications.

2. Considerable technology is required to recover the solid polymer.

3. The water in the reaction mass lowers the yield per reactor volume.

REFERENCES

1. Flory, P. J. *Principles of Polymer Chemistry.* Cornell University Press, Ithaca, New York, 1953.

2. Schildknecht, C. E. *Polymer Processes.* p. 38. John Wiley & Sons, Inc. New York, 1956.

3. Wohl, M. H. Bulk polymerization. *Chem. Eng.*, Aug. 1, p. 60, 1962.

4. Sittig, M. Polyolefin processes today. *Petrol. Refiner*, **39,** no. 11, p. 162, 1960.

5. Sittig, M. How to make polypropylene. *Petrol. Refiner*, **40,** no. 3, p. 129, 1961.

6. Anspon, H. D. *Manufacture of Plastics.* (W. M. Smith, ed.), ch. 2. Reinhold Publishing Co., New York, 1964.

7. Jezl, J. E., et. al. *Ibid*, ch. 3.

8. Schildknecht, C. E. *Op. cit.*, p. 94.

9. Guccione, E. New developments in acrylic processing. *Chem. Eng.*, June 6, p. 138, 1966.

10. Bovey, F. A. *et al. Emulsion Polymerization.* John Wiley & Sons, Inc. New York, 1955.

11. Flory, P. J. *Principles of Polymer Chemistry.* Ch. V-3. Cornell University Press, Ithaca, New York, 1953.

12. Harkins, W. D. A general theory of the mechanism of emulsion polymerization. *J. Amer. Chem. Soc.*, **69**, p. 1428, 1947.

Section III

MECHANICAL PROPERTIES
OF POLYMERS

14
Rubber Elasticity

14.1 INTRODUCTION

Natural and synthetic rubbers possess some interesting, unique and useful mechanical properties. No other materials are capable of reversible extension to strains of 6–700 percent. No other materials exhibit an increase in modulus with increasing temperature. It was recognized long ago that vulcanization was necessary in order that rubber deformation be completely reversible. We now know that this is a result of the crosslinks so introduced, preventing the bulk slippage of the molecules past one another and eliminating flow (irrecoverable deformation). Thus, when a stress is applied to a sample of crosslinked rubber, equilibrium is established fairly rapidly. Once at equilibrium, the properties of the rubber can be described by thermodynamics.

14.2 THERMODYNAMICS OF ELASTICITY

Consider an element of material with dimensions $a \times b \times c$, as sketched in Figure 14.1. By applying the first law of thermodynamics to this system,

$$dU = dQ - dW \qquad (14\text{-}1)$$

where dU is the change in the system's *internal energy* and dQ and dW are the heat and work exchanged between system and surroundings as the system undergoes a differential change. (We have adopted the convention here that work done *by the system on the surroundings* is positive.)

153

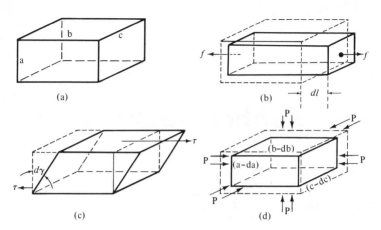

Figure 14.1. Types of mechanical deformation. (a) unstressed, (b) pure tension, (c) pure shear, (d) isotropic compression.

In general, there are three types of mechanical work possible:
a. work done by a tensile force f:

$$dW(\text{tensile}) = -f\,dl \qquad (14\text{-}2)$$

where dl is the differential change in the system's length arising from the application of the force f. This is the fundamental definition of work. The negative sign arises from the need to reconcile the mechanical convention of treating a tensile force (which does work *on* the system) as positive with the thermodynamic convention above.

b. work done by a shear stress τ:

$$dW(\text{shear}) = (\text{force})(\text{distance}) = -(\tau\,bc)(a\,d\gamma) = -\tau V d\gamma \qquad (14\text{-}3)$$

where γ is the shear strain (Figure 14.1c) and $V = abc =$ the system volume.

c. work done by an isotropic pressure in changing the volume:

$$dW(\text{pressure}) = P(cb)da + P(ac)db + P(ab)dc = PdV. \qquad (14\text{-}4)$$

Note that no minus sign is needed here. A positive pressure causes a decrease in volume (negative dV) and does work *on* the system.

If the deformation process is assumed to occur *reversibly* (in a thermodynamic sense),

$$dQ = TdS \tag{14-5}$$

where S is the system's entropy.

Combining the preceding five equations gives a general relation for the change of internal energy of an element of material undergoing a differential deformation:

$$dU = TdS - PdV + fdl + V\tau d\gamma. \tag{14-6}$$

Now, let's consider three individual types of deformation:

a. Pure tension at constant volume and temperature. Under these conditions, $dV = \tau = 0$. Dividing the remaining terms in (14-6) by dl, restricting to constant T and V, and solving for f gives:

$$f = \left(\frac{\partial U}{\partial l}\right)_{T,V} - T\left(\frac{\partial S}{\partial l}\right)_{T,V} \tag{14-7}$$

b. Pure shear at constant volume and temperature. Here, $dV = f = 0$. Dividing the remaining terms in (14-6) by $d\gamma$, restricting to constant T and V, and solving for τ gives

$$\tau = \frac{1}{V}\left(\frac{\partial U}{\partial \gamma}\right)_{T,V} - \frac{T}{V}\left(\frac{\partial S}{\partial \gamma}\right)_{T,V}. \tag{14-8}$$

c. Isotropic compression only, at constant temperature, is

$$P = \left(\frac{\partial U}{\partial V}\right)_T + T\left(\frac{\partial S}{\partial V}\right)_T. \tag{14-9}$$

It must be mentioned here that, from an experimental standpoint, it is very difficult to carry out tensile experiments at constant volume to obtain the partial derivatives in (14-7). Most tests are carried out at constant pressure (atmospheric), and, in general, there is a change in volume with tensile straining. Fortunately, Poisson's ratio is approximately 0.5 for rubbers, so this change in volume is small, and (14-7) is approximately valid for tensile deformation at constant pressure also. For precise work, the hydrostatic pressure must be varied to maintain V constant, or corrections applied to the constant-pressure data to obtain the constant-volume coefficients (*1*, *2*). In pure shear, *V* should be constant and (14-8) should be valid.

Types of Elasticity   Equations (14-7), (14-8), and (14-9) reveal that there are *energy* (the first term on the right) and *entropy* (the second term on the right) contributions to the tensile force, shear stress or isotropic pressure.   In polymers, *energy elasticity* represents the storage of energy resulting from the elastic straining of bond angles and lengths like springs from their equilibrium values.   *Entropy elasticity* is caused by the decrease in entropy upon straining.   This can be visualized by considering a single polymer molecule subjected to a tensile stress.   In an unstressed state, it is free to adopt an extremely large number of random, balled-up configurations (Figure 14.2a), switching from one to another through rotation

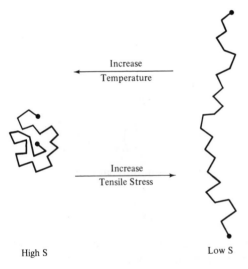

Figure 14.2.   The effects of stress and temperature on chain configurations.

about bond angles. Now imagine it to be stretched out under the application of a tensile force (Figure 14.2b). It is obvious that there are far fewer configurational possibilities. The more it is stretched, the fewer they become. Now $S = k \ln \Omega$, where k is Boltzmann's constant and Ω is the number of configurational possibilities, so *stretching decreases the entropy* (increases the order). Raising the temperature has precisely the opposite effect. The added thermal energy of the chain segments increases the intensity of their lateral vibrations, favoring a return to the more random or higher entropy state. This tends to pull the extended chain ends together, giving rise to a retractive force.

The Ideal Rubber   In a gas subjected to an isotropic pressure, the energy term in (14-9) arises from the change in intermolecular forces with volume, and the entropy term arises from the increased room (and therefore greater disorder) the molecules gain with volume. In an *ideal gas*, there are *no* intermolecular forces, $(\partial U/\partial V)_T = 0$. By analogy, in an *ideal rubber*, $(\partial U/\partial l)_{T,V} = (\partial U/\partial \gamma)_{T,V} = 0$, and elasticity arises only from entropy effects. For many gases around room temperature and above, and around atmospheric pressure and below, $(\partial U/\partial V)_T < T(\partial S/\partial V)_T$, and the ideal gas law is a good approximation. Similarly, as illustrated in Figure 14.3,

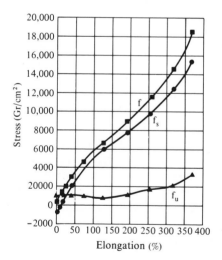

Figure 14.3.   Energy (f_u) and entropy (f_s) contributions to tensile stress in natural rubber at 20C (*3*). (Copyright 1942 by the American Chemical Society. Reprinted by permission of the copyright owner.)

$(\partial U/\partial l)_{T,V} < T(\partial S/\partial l)_{T,V}$ for rubbers, and they behave approximately as ideal rubbers.

Effects of Temperature at Constant Force

Now let's consider what happens to the length of a piece of rubber when its temperature is changed while a weight is suspended from it—i.e., while it is maintained at a constant tensile force. Assuming constant volume (an approximation, in the usual constant-pressure experiment), $dU = TdS + fdl$. Solve for l, divide by dT, and restrict to constant f as well as V; then

$$\left(\frac{\partial l}{\partial T}\right)_{f,V} = \frac{1}{f}\left(\frac{\partial U}{\partial T}\right)_{f,V} - \frac{T}{f}\left(\frac{\partial S}{\partial T}\right)_{f,V}. \qquad (14\text{-}10)$$

As before, the first term on the right represents energy elasticity, and the second represents entropy elasticity. Since internal energy generally increases with temperature, the partial derivative in the energy term is positive, as is f. The energy term, therefore, causes an increase in length with temperature, (positive $(\partial l/\partial T)_{f,V}$). This is the normal thermal expansion, observed in all materials (metals, ceramics, glasses, etc.), reflecting the increase in the average distance between molecular centers with temperature. All factors in the entropy term are positive, however, and, since it is preceded by the negative sign, it gives rise to a *decrease* in length with increasing temperature. In rubbers, where the entropy effect overwhelms the normal thermal expansion, this is what is actually observed. In all other materials, where the structural units are confined to a single arrangement (e.g., the atoms in a crystal lattice cannot readily interchange) the entropy term is negligible. The magnitude of the entropy contraction in rubbers is much greater than the thermal expansion of other materials. An ordinary rubber band will contract an inch or so when heated to 300F under stress, while the expansion of a piece of metal of similar length over a similar temperature range would not be noticeable to the naked eye.

Effects of Temperature at Constant Length

It is interesting to consider what happens to the force in a piece of rubber when it is heated while stretched to a constant length.

By using the exact thermodynamic Maxwell relation

$$-\left(\frac{\partial S}{\partial l}\right)_{T,V} = \left(\frac{\partial f}{\partial T}\right)_{L,V} \qquad (14\text{-}11a)$$

to describe approximately the usual experiment conducted at constant pressure (a better approximation to the constant pressure experiment is

$$-\left(\frac{\partial S}{\partial l}\right)_{T,V} \simeq \left(\frac{\partial f}{\partial T}\right)_{P,\alpha} \qquad (14\text{-}11b)$$

where α, the extension ratio $= l/l_o$, the ratio of stretched to unstretched length at a particular temperature), and combining (14-11a) and (14-7) gives

$$\left(\frac{\partial f}{\partial T}\right)_{L,V} = \frac{f}{T} - \frac{1}{T}\left(\frac{\partial U}{\partial l}\right)_{T,V}. \qquad (14\text{-}12)$$

Now, both f and T are positive, so the first term on the right causes the force to increase with temperature, a result of the greater thermal agitation (tendency toward higher entropy) of the extended chains. The partial derivative in the second term is also positive, as energy is stored springlike in the strained bond angles and lengths with extension. With the negative sign in front, this term predicts a relaxation of the tensile force with increasing temperature. Again, this second term reflects the ordinary thermal expansion obtained with all materials, but in rubbers, at reasonably large values of f, it is overshadowed by the first (entropy) term, and the force increases with temperature. For an *ideal* rubber, $(\partial U/\partial l)_{T,V} = 0$, and integration of (14-12) at constant volume gives

$$f = (\text{constant})T \quad (\text{ideal rubber}). \qquad (14\text{-}13)$$

This is analogous to the linearity between P and T in an ideal gas at constant V. These observations are confirmed in Figure 14.4. The negative slope at low elongations arises from the predominance of thermal expansion when elongation, and hence f, is low. Note that there is an intermediate elongation, the *thermoelastic inversion* point, at which force will be essentially independent of temperature, where thermal expansion and entropy contraction balance.

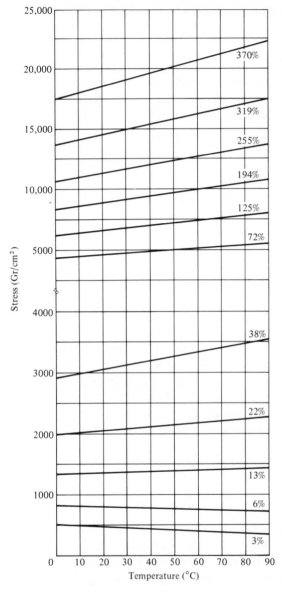

Figure 14.4.   Force vs. temperature in natural rubber maintained
at constant extension (% relative to length at 20C) (3).
(Copyright 1942 by the American Chemical Society.
Reprinted by permission of the copyright owner.)

14.3 STATISTICS OF IDEAL RUBBER ELASTICITY (1, 4, 5)

A typical rubber consists of long chains connected by short crosslinks every few hundred carbon atoms. The chain segments between crosslinks are known as *network chains*. The change in entropy upon stretching a sample containing N moles of network chains is

$$S - S_o = NR \ln \Omega/\Omega_o \qquad (14\text{-}14)$$

where the subscript o refers to the unstretched state, Ω is the number of configurations available to the N moles of network chains and R is the gas constant. By statistically evaluating the Ω's, it is possible to show that, for constant-volume stretching,

$$S - S_o = -\frac{1}{2} NR \left[\left(\frac{l}{l_o} \right)^2 + 2 \left(\frac{l_o}{l} \right) - 3 \right]. \qquad (14\text{-}15)$$

For an *ideal rubber*, in which the tensile force is given by

$$f = -T \left(\frac{\partial S}{\partial l} \right)_{T,V} \quad \text{(ideal rubber)}, \qquad (14\text{-}16)$$

differentiation of (14-15) and insertion into (14-16) then gives

$$f = \frac{NRT}{l_o} \left[\left(\frac{l}{l_o} \right) - \left(\frac{l_o}{l} \right)^2 \right]. \qquad (14\text{-}17)$$

However,

$$N = \frac{\text{mass}}{\overline{M}_c} = \frac{\rho V}{\overline{M}_c} = \frac{\rho l_o A_o}{\overline{M}_c} = \frac{\rho l A}{\overline{M}_c} \qquad (14\text{-}18)$$

where $\overline{M}_c$ is the number-average molecular weight of the network chains, ρ the density, and A the cross-sectional area of the sample (since the volume change in stretching a piece of rubber is negligible, $A_o l_o = Al$). Therefore,

$$f = \frac{\rho A_o RT}{\overline{M}_c} \left[\left(\frac{l}{l_o} \right) - \left(\frac{l_o}{l} \right)^2 \right]. \qquad (14\text{-}19)$$

The *engineering tensile stress* is defined as the tensile force divided by the initial cross-sectional area of the sample and is, therefore,

$$\sigma \text{ (engineering)} = \frac{f}{A_o} = \frac{\rho RT}{\overline{M}_c}\left[\left(\frac{l}{l_o}\right) - \left(\frac{l_o}{l}\right)^2\right] \quad (14\text{-}20)$$

and the *true tensile stress*, defined as the tensile force over the actual area A at length l, is

$$\sigma \text{ (true)} = \frac{f}{A} = \frac{\rho RT}{\overline{M}_c}\left[\left(\frac{l}{l_o}\right)^2 - \left(\frac{l_o}{l}\right)\right]. \quad (14\text{-}21)$$

Since the tensile strain ϵ is

$$\epsilon = (l - l_o)/l_o, \quad (14\text{-}22)$$

the slope of the true stress-strain curve (tangent Young's modulus) is

$$E = \left(\frac{\partial \sigma}{\partial \epsilon}\right) = \left(\frac{\partial \sigma}{\partial l}\right)\left(\frac{\partial l}{\partial \epsilon}\right) = \frac{\rho RT}{\overline{M}_c}\left[2\left(\frac{l}{l_o}\right) + \left(\frac{l_o}{l}\right)\right] \quad (14\text{-}23)$$

and the initial modulus (as $l \rightarrow l_o$) for an ideal rubber becomes

$$E \text{ (initial)} = \frac{3\rho RT}{\overline{M}_c}. \quad (14\text{-}24)$$

Equations (14-19) through (14-24) point out two important facts: first, that the force (or modulus) in an ideal rubber sample held at a particular strain increases in proportion to the absolute temperature (confirming 14-13), and second, that it is *inversely* proportional to the molecular weight of the chain segments between crosslinks. Thus, increased crosslinking, which reduces $\overline{M}_c$, is an effective means of stiffening a rubber. Equation (14-24) is often used to obtain $\overline{M}_c$ from mechanical tests and thereby evaluate the efficiency of various crosslinking procedures. Even noncrosslinked polymers exhibit rubbery behavior above their T_g's for short periods of time. This is due to mechanical entanglements acting as temporary crosslinks, and $\overline{M}_c$ then represents the average length of chain segments between entanglements.

When compared with experimental data, (14-19) does a good job in compression, but it begins to fail at extension ratios

(l/l_o) greater than about 1.5 where the experimental force becomes greater than predicted. There are a number of reasons for this. First, (14-15) is based on the assumption of a Gaussian distribution of network chains. This assumption fails at high elongations, and it is also in error if crosslinks are formed when chains are in a strained configuration. Second, it does not take into account the presence of chain end segments, which do not contribute to the support of stress. Third, rubbers (natural rubber in particular) may begin to crystallize as a result of chain orientation at high elongations. This causes the stress-strain curve to shoot up markedly. Theoretical modifications are available for the first two factors which improve things considerably in the absence of crystallinity, but there is as yet no satisfactory treatment of the effect of crystallinity on stress-strain properties.

It is important to keep in mind also that, in practice, rubbers are rarely used in the form of pure polymer. They almost always are reinforced with carbon black and often contain other fillers, plasticizing and extending oils, etc., all of which influence the stress-strain properties and are not considered by the theories discussed.

REFERENCES

1. Flory, P. J. *Principles of Polymer Chemistry*, ch. XI. Cornell University Press, Ithaca, New York, 1953.

2. Meares, P. *Polymers: Structure and Bulk Properties*, ch. 6. Van Nostrand Co., Inc., Princeton, N. J., 1965.

3. Anthony, R. L., R. H. Caston and E. Guth, Equations of state for natural and synthetic rubber-like materials. *J. Phys. Chem.*, **46,** no. 826, 1942.

4. Tobolsky, A. V. *Properties and Structure of Polymers*, ch. II. John Wiley & Sons, Inc., New York, 1960.

5. Meares. Op. cit., ch. 7 and 8.

15
Purely Viscous Flow

15.1 INTRODUCTION

Rheology (the study of deformation and flow) is currently one of the most intense areas of research interest. Much rheological work is concerned directly with polymers simply because they exhibit such interesting, unusual and difficult-to-describe (at least from the standpoint of traditional materials) deformation behavior. The simple and traditional linear engineering models, Newton's law (for flow) and Hooke's law (for elasticity) just aren't often reasonable approximations. Not only are the elastic and viscous properties of polymer melts and solutions usually nonlinear but they exhibit a *combination* of viscous and elastic response, the relative magnitudes of which depend on the time scale of the experiment. This *viscoelastic* response is dramatically illustrated by 'silly putty' (a silicone polymer). When bounced (stressed rapidly), it is highly elastic, recovering most of the potential energy it had before being dropped. If stuck on the wall, however, (stressed over a long time period) it will slowly flow down the wall, albeit with a high viscosity, and will show little tendency to recover any deformation.

15.2 BASIC DEFINITIONS

We begin our treatment of rheology with a discussion of *purely viscous flow*. For our purposes, this will be defined as a deformation process in which all the applied mechanical energy is nonrecoverably dissipated as heat in the material; i.e., it is all converted to heat through *viscous energy dissipation*. Purely viscous flow is in most cases a good approximation for dilute polymer solutions and often is for concentrated solutions and melts where the stress on the material is not changing too rapidly—i.e., where it has a chance to approach an equilibrium flow situation.

The *viscosity* of a material expresses its resistance to flow under a mechanical stress. It is defined quantitatively in terms of two basic parameters, the *shear stress*, τ, and the *shear rate* (more correctly, the rate of shear straining), $\dot{\gamma}$. These quantities are defined in Figure 15.1. Consider a point in a laminar

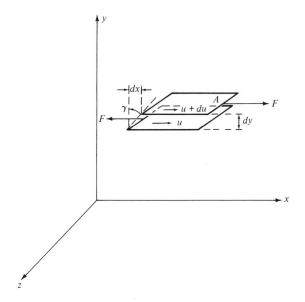

Figure 15.1. Definitions of shear stress and shear rate.

flow field (Figure 15.1). (We limit ourselves here to two dimensions. A detailed three-dimensional treatment of rheology is beyond the scope of this book, but several excellent treatises are available (*1, 2*).) A coordinate system is established such that the x-axis (sometimes designated the 1 coordinate direction) is in the direction of flow, and the y-axis (the 2 direction) is perpendicular to surfaces of constant fluid velocity; i.e., it is parallel to the *velocity gradient*. The z-axis (3 or neutral direction) is mutually perpendicular to the others. A flow field

in which the velocity and its gradient are everywhere perpendicular is known as a *viscometric* or *simple shearing flow* and,
from a practical standpoint, is one type which can be simply
treated analytically. Fortunately, many laminar-flow situations encountered are, or at least can reasonably be approximated by viscometric flows. Examples of viscometric flows are
given in chapter 16.

A fluid surface at y moves with a velocity $u = dx/dt$ in the
x direction, while the surface at $y + dy$ has a velocity $u + du$.
The displacement gradient, dx/dy, is known as the *shear strain*
and is given the symbol γ.

$$\gamma = dx/dy = \text{shear strain (dimensionless)} \qquad (15\text{-}1)$$

The *time rate of change* of shear strain, $\dot\gamma$, (the dot denotes the
time derivative) is the so-called *shear rate*. Since the order in
which the mixed second derivative is taken is immaterial (note
that, by sticking to two dimensions, we can write total rather
than partial derivatives),

$$\dot\gamma = \left(\frac{d}{dt}\right)(\gamma) = \frac{d}{dt}\left(\frac{dx}{dy}\right) = \frac{d}{dy}\left(\frac{dx}{dt}\right) = \frac{du}{dy}\ (\text{time}^{-1}). \quad (15\text{-}2)$$

Thus, an alternate definition of the shear rate is the velocity
gradient, du/dy.

The shear stress is simply the force (in the direction of flow)
per unit area normal to the y axis

$$\tau_{yx} = \frac{F\,(\text{in } x\text{-direction})}{A\,(\text{normal to } y\text{-direction})}\left(\frac{\text{force}}{\text{length}^2}\right). \qquad (15\text{-}3)$$

The subscript yx will henceforth be dropped unless specifically
needed.

15.3 RELATIONS BETWEEN τ AND $\dot\gamma$—FLOW CURVES

Newton's law of viscosity states that the shear stress is
linearly proportional to the shear rate, the proportionality
constant being the *viscosity*, η.

$$\tau = \eta\dot\gamma \qquad (15\text{-}4)$$

(Most rheological work is done in the cgs system with force in dynes, mass in grams, length in centimeters, and time in seconds. In this system, the unit of viscosity is dyne-sec/cm^2 or the *poise*. All equations here will be written with this system in mind. When using the English system of pound force, pound mass, feet and seconds, each stress or pressure as written here must be multiplied by the dimensional *constant* g_c = 32.2 ft-lb$_m$/lb$_f$-sec^2.) Fluids which obey this hypothesis are termed Newtonian. It holds quite well for low molecular weight fluids such as gases, water, motor oil, etc. An arithmetic plot of τ versus $\dot\gamma$, a *flow curve*, is a straight line through the origin with a slope η for a Newtonian fluid (Figure 15-2a). By taking

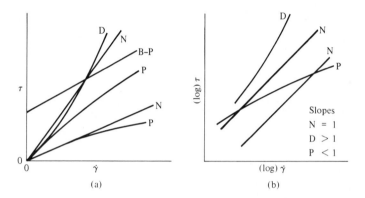

Figure 15.2.   Types of flow curves N = Newtonian, P = pseudo-plastic, B-P = Bingham plastic (infinitely pseudo-plastic), D = dilatant. (a) arithmetic, (b) logarithmic.

logarithms of both sides of (15-4),

$$\log \tau = \log \eta + 1 \log \dot\gamma.$$

Hence, a log-log plot of τ versus $\dot\gamma$, a logarithmic flow curve, will be a line of slope unity for a Newtonian fluid (Figure 15.2b). This type of flow behavior would be expected for small molecules where the structure, and therefore the resistance to flow, does not change with the intensity of shearing.

Unfortunately, many fluids do not obey Newton's hypothesis. Both *dilatant* (shear-thickening) and *pseudoplastic* (shear-

thinning) fluids have been observed (Figure 15.2). On logarithmic coordinates, dilatant flow curves have a slope greater
than 1 and pseudoplastics less than 1. Dilatant behavior is
reported for certain slurries and implies an increased resistance
to flow with intensified shearing. *Polymer melts and solutions
are invariably pseudoplastic*; i.e., their resistance to flow decreases with the intensity of shearing.

An entirely equivalent method of representing a material's
equilibrium viscous properties is the *apparent viscosity*, η_a.

$$\eta_a = \frac{\tau}{\dot{\gamma}} \qquad (15\text{-}5)$$

For a Newtonian fluid, $\eta_a = \eta\dot{\gamma}/\dot{\gamma} = \eta$, a constant. For non-
Newtonian fluids, of course, the apparent viscosity is a func-
tion of shear rate (or shear stress). A knowledge of the relation
between any two of the three variables η_a, τ, and $\dot{\gamma}$ completely
defines the equilibrium viscous properties, since they are re-
lated by (15-5).

15.4 TIME-DEPENDENT BEHAVIOR

The types of non-Newtonian flow just described, though
shear dependent, are time independent; i.e., as long as a constant shear rate or stress is maintained, the same apparent viscosity will be observed at equilibrium. Some fluids exhibit
reversible time-dependent properties, however. When sheared
at a constant rate or stress, the apparent viscosity of a *thixo-
tropic* fluid will decrease over a period of time (Figure 15.3),
implying a progressive breakdown of structure. If the shearing
is stopped for a while, the structure reforms, and the experiment may be duplicated. The ketchup which splashes all over
after a period of vigorous tapping is a classic example. Thixo-
tropic behavior is important in the paint industry, where
smooth, even application with brush or roller is required, but
it is desirable for the paint on the surface to 'set up' to avoid
drips and runs after application. The opposite sort of behavior
is manifested by *rheopectic* fluids—e.g., certain drilling muds
used by the petroleum industry. When subjected to continuously increasing and then decreasing shearing, time-dependent
fluids give flow curves as in Figure 15.4.

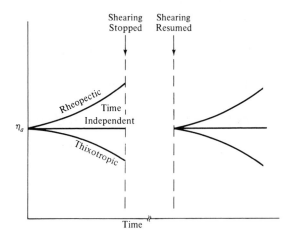

Figure 15.3.   Time-dependent fluids.

Once in a while, polymer systems will appear to be thixo-tropic or rheopectic. Careful checking (including before and after molecular-weight determination) invariably shows that the phenomenon is not reversible and is due to degradation or crosslinking of the polymer when in the viscometer for long

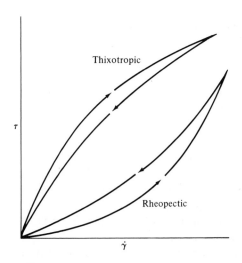

Figure 15.4.   Flow curves for time-dependent fluids under continu-ously increasing then decreasing shear.

periods of time, particularly at elevated temperatures. Other *transient* time-dependent effects in polymers are due to elasticity and will be considered later, but, for chemically stable polymer systems, the *equilibrium* viscous properties are time independent. We treat only such systems from here on.

15.5 POLYMER MELTS AND SOLUTIONS

When the flow properties of polymer melts and solutions can be measured over a wide enough range of shearing, the logarithmic flow curves appear as in Figure 15.5a. It is gen-

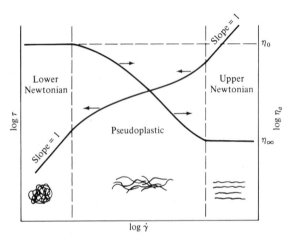

Figure 15.5. Generalized flow properties for polymer melts and solutions.

erally observed that

1. at low shear rates (or stresses), a lower Newtonian region is reached with a so-called *zero-shear viscosity*, η_o;

2. over several decades of intermediate shear rates, the material is pseudoplastic; and

3. at very high shear rates, an upper Newtonian region with viscosity η_∞ is observed.

This behavior can be rationalized in terms of molecular structure. At low shear, the randomizing effect of the thermal motion of the chain segments overcomes any tendency toward molecular alignment in the shear field. The molecules are thus in their most random and highly entangled state and have their greatest resistance to slippage (flow). As the shear is increased, the molecules will begin to untangle and align in the shear field, reducing their resistance to slippage past one another. Under severe shearing, they will be pretty much completely

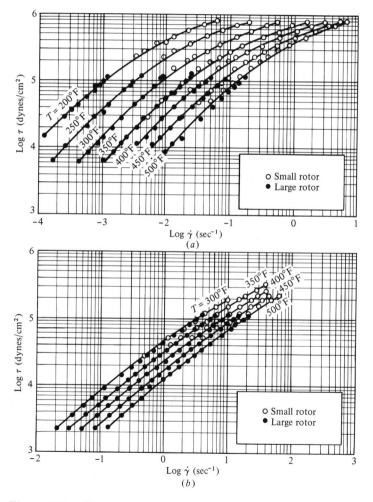

Figure 15.6.   Flow curves for polymer melts (3); (a) L-80 polyisobutylene, (b) low-density polyethylene.

untangled and aligned and reach a state of minimum resistance to flow. This is illustrated schematically in Figure 15.5b. Intense shearing will eventually lead to extensive breakage of main-chain bonds—i.e., mechanical degradation.

It is worthwhile to consider here what happens to the highly oriented molecules when the shear field is removed. The randomizing effect of thermal energy tends to return them to their low-shear configurations, giving rise to an elastic retraction.

Some actual flow curves for polymer melts are shown in Figure 15.6 (3). The data cover only a portion of the general range described above.

15.6 QUANTITATIVE REPRESENTATION OF FLOW BEHAVIOR

In order to handle non-Newtonian flow analytically, it is desirable to have a mathematical expression relating τ and $\dot{\gamma}$ as Newton's law does for Newtonian fluids. A wide variety of such *constitutive relations* has been proposed, both theoretical and empirical (4, 5). All appear to fit at least some experimental data over a limited range of shear rates but, in general, the more adjustable parameters there are in the equation, the better fit it provides. (There's an old saying that with six constants you can draw an elephant and, with a seventh, make his trunk wave.) The mathematical complexity of the equations increases greatly with the number of parameters, soon outstripping the data available to establish the parameters and making the equations impractical for engineering calculations.

The most common engineering model for purely viscous non-Newtonian flow is the so-called *power law*

$$\tau = K(\dot{\gamma})^n. \qquad (15\text{-}6a)$$

This is a two-parameter model with the adjustable parameters being K, the *consistency*, and n, the *flow index*. As written above, the dimensions of K depend on the magnitude of n (dimensionless), so, strictly speaking, the power law should be (but rarely is) written

$$\tau = K|\dot{\gamma}|^{n-1}\dot{\gamma}. \qquad (15\text{-}6b)$$

This way, K has the usual viscosity units.

On log τ versus log $\dot\gamma$ coordinates, a power-law fluid is represented by a straight line with slope n. Thus, for $n = 1$, it reduces to Newton's law; for $n < 1$, the fluid is pseudo-plastic; and for $n > 1$, dilatant. It can reasonably approximate only portions of actual flow curves over one or two decades of shear rate (see Figure 15.6), but it does so with fair mathematical simplicity and is adequate for many engineering situations. Many useful relations are obtained simply by replacing Newton's law with the power law in the usual fluid-dynamic equations.

Example 1. Determine the equation relating apparent viscosity to shear rate for a power-law fluid.

Solution.

$$\eta_a = \frac{\tau}{\dot\gamma} = \frac{K\dot\gamma^n}{\dot\gamma} = K(\dot\gamma)^{n-1} \tag{15-7}$$

Thus, a log-log plot of η_a versus $\dot\gamma$ for a power-law fluid will be linear with a slope of $(n - 1)$. This points up an interesting limitation of the power law. Where the shear rate goes to zero (e.g., for Poiseuille flow at the center line of a cylindrical tube), the apparent viscosity approaches infinity for a pseudoplastic fluid.

15.7 TEMPERATURE DEPENDENCE OF FLOW PROPERTIES

Also of engineering interest is the variation of flow properties with temperature. The *zero-shear* viscosity can often be represented by the relation

$$\eta_o = Ae^{E/RT} \tag{15-8}$$

over a range of several hundred °F above T_g, as can the viscosity of low molecular weight fluids, but polymer systems are rarely in the lower Newtonian range in commercial processing operations. Since the apparent viscosity is a function of temperature and shear stress *or* shear rate,

$$\eta_a = f(\tau, T) \quad or \quad \eta_a = f'(\dot\gamma, T).$$

Over a similar temperature range, these functions are approximated by

$$\eta_a = Be^{E_\tau/RT} \quad (\tau \text{ constant}) \tag{15-9a}$$

$$\eta_a = Ce^{E_{\dot\gamma}/RT} \quad (\dot\gamma \text{ constant}) \tag{15-9b}$$

where E_τ is the activation energy for flow at constant shear stress and $E_{\dot\gamma}$ is the activation energy for flow at constant shear rate. Figure 15.7 shows plots of log $\dot\gamma$ versus $1/T$ at constant τ(3). Since $\dot\gamma = \tau/\eta_a$, the slope of these plots is $-E_\tau/R$.

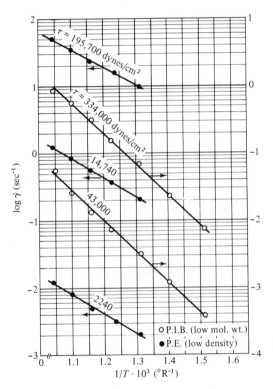

Figure 15.7.   Temperature dependence of shear rate at constant shear stress (3).

Example 2. What must the temperature be to halve the viscosity of the polymers in Figure 15.7 at 300F? The activation energies E_τ are 8,830 cal/g mole for the polyisobutylene (PIB) and 5,580 cal/g mole for the polyethylene (PE).

Solution. At any value of shear stress τ, the apparent viscosity is given by (15-9a). Therefore, between temperatures T_1 and T_2

$$\frac{\eta_{a_2}}{\eta_{a_1}} = \frac{e^{E_\tau/RT_2}}{e^{E_\tau/RT_1}} = e^{(E_\tau/R)(1/T_2 - 1/T_1)}$$

or $\ln \dfrac{\eta_{a_2}}{\eta_{a_1}} = \dfrac{E_\tau}{R} \left(\dfrac{1}{T_2} - \dfrac{1}{T_1} \right).$

In this case,

$$\frac{\eta_{a_2}}{\eta_{a_1}} = \frac{1}{2}, \qquad \ln \frac{1}{2} = -0.693.$$

For PIB,

$$\frac{1}{T_2} - \frac{1}{T_1} = \frac{(-0.693)(1.99 \text{ cal/g mol } ^\circ K)}{8,830 \text{ cal/g mole}}$$

$$= -0.000156 \left(\frac{1}{^\circ K} \right)$$

$$= -0.000087 \left(\frac{1}{^\circ R} \right)$$

$$\frac{1}{T_1} = \frac{1}{460 + 300} = 0.001316 \left(\frac{1}{^\circ R} \right)$$

$$\frac{1}{T_2} = 0.001316 - 0.000087 = 0.001229 \left(\frac{1}{^\circ R} \right)$$

$$T_2 = 813 R = 353 F$$

for PE

$$\frac{1}{T_2} - \frac{1}{T_1} = \frac{(-0.693)(1.99)}{5,580} = -0.000247 \left(\frac{1}{^\circ K} \right)$$

$$= -0.00137 \left(\frac{1}{^\circ R} \right)$$

$$\frac{1}{T_2} = 0.001316 - 0.000137 = 0.001179 \left(\frac{1}{^\circ R} \right)$$

$$T_2 = 848 R = 388 F$$

The temperature is, therefore, an effective means of controlling melt viscosity in processing operations, but two drawbacks must be kept in mind: (a) it takes time and costs money to put in and take out thermal energy, and (b) excessive temperatures can lead to degradation of the polymer.

A more general expression for the variation of η_o with temperature is obtained from the principle of corresponding states, which says that all amorphous polymers are mechanically equivalent equally removed from their glass transition temperatures. It is expressed quantitatively by the famous Williams-Landel-Ferry or *WLF* equation:

$$\log \frac{\eta_o(T)}{\eta_o(T_g)} = -\frac{17.44\,(T - T_g)}{51.6 + T - T_g}. \tag{15-10}$$

Modifications of (15-10) have been proposed which use different constants and characteristic temperatures other than $T_g(6)$. However, because T_g is extensively tabulated, (15-10) is of most practical value and does a pretty good job of representing experimental results for a wide variety of polymers in the temperature range of T_g to $T_g + 100C$.

15.8 INFLUENCE OF MOLECULAR WEIGHT ON FLOW PROPERTIES

It has long been known that a polymer's molecular weight exerts a strong influence on the melt or concentrated solution viscosity. Experiments show that

$$\eta_o \propto (\overline{M}_w)^1 \qquad \text{for } \overline{M}_w < \overline{M}_{w_c} \tag{15-11a}$$

$$\eta_o \propto (\overline{M}_w)^{3.4} \qquad \text{for } \overline{M}_w > \overline{M}_{w_c} \tag{15-11b}$$

where $\overline{M}_{w_c}$ is a critical average molecular weight, thought to be the point at which molecular entanglements begin to dominate the rate of slippage of the molecules. It depends on the polymer type, but most commercial polymers are well above $\overline{M}_{w_c}$.

Equation (15-11) holds quantitatively for just about all polymer melts. The addition of a low molecular weight solvent, of course, cuts down entanglements and raises $\overline{M}_{w_c}$. Nevertheless, even moderately concentrated (say 25 percent or more) polymer solutions have viscosities proportional to $\overline{M}_w^{3.4}$ provided $\overline{M}_w$ is in the range of commercial importance. As the shear rate is increased, the number of entanglements between chains is reduced, and, as expected, the dependence of viscosity on molecular weight decreases (Figure 15.8).

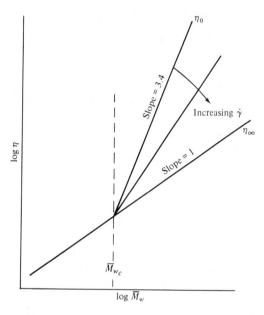

Figure 15.8. The effect of molecular weight on viscosity.

Example 3. By what percentage must $\overline{M}_w$ be changed to cut η_o in half?

Solution. From (15-11b),

$$\frac{\eta_{o_2}}{\eta_{o_1}} = \left(\frac{\overline{M}_{w_2}}{\overline{M}_{w_1}}\right)^{3.4}$$

$$\frac{\overline{M}_{w_2}}{\overline{M}_{w_1}} = \left(\frac{\eta_{o_2}}{\eta_{o_1}}\right)^{1/3.4} = \left(\frac{1}{2}\right)^{1/3.4} = 0.816$$

$$\text{percent change} = 100\left(\frac{\overline{M}_{w_2} - \overline{M}_{w_1}}{\overline{M}_w}\right) = 100\left(\frac{\overline{M}_{w_2}}{\overline{M}_{w_1}} - 1\right)$$

$$= -18.4 \text{ percent}$$

Thus, it only takes an 18 percent decrease in chain length to halve the melt viscosity. Although things won't be quite so dramatic at high shear rates, this illustrates the importance of controlling molecular weight to achieve the desired processing properties.

15.9 THE EFFECTS OF PRESSURE ON VISCOSITY

For all fluids, viscosity increases with increasing pressure as the free volume, and hence the case of molecular slippage, is decreased. With liquids (including polymer melts), because of their relative incompressibility, the effect becomes noticeable only at very high pressures (many thousand psi). The available data are fitted by an equation of the type

$$\eta = Ae^{BP} \tag{15-12}$$

where A and B are constants.

Carley (7) has critically reviewed work on polymer melts and concludes that pressure effects are of minor significance in most processing situations, provided the temperature is not too close to a transition. High pressures raise both T_g and T_m slightly, and of course the viscosity shoots up tremendously as either is reached.

REFERENCES

1. Coleman, B. D., H. Markovitz and W. Noll. *Viscometric Flows of Non-Newtonian Fluids.* Springer-Verlag Inc., New York, 1966.

2. Middleman, S. *The Flow of High Polymers.* John Wiley & Sons, Inc., New York, 1968.

3. Best, D. M. and S. L. Rosen. A simple, versatile and inexpensive rheometer for polymer melts. *Polymer Eng. and Sci.*, **8,** no. 2, p. 116, 1968.

4. Bird, R. B., W. E. Stewart and E. N. Lightfoot. *Transport Phenomena*, ch. 1. John Wiley & Sons, Inc., New York, 1960.

5. Rodriguez, F. *Principles of Polymer Systems*, ch. 7. McGraw-Hill Book Co., New York, 1970.

6. Tobolsky, A. V. *Properties and Structure of Polymers*, ch. II.8. John Wiley & Sons, Inc., New York, 1960.

7. Carley, J. F. Effect of static pressure on polymer melt viscosities. *Mod. Plast.*, **39,** no. 4, p. 123, 1961.

16
Viscometry and Tube Flow

16.1 INTRODUCTION

This chapter will consider some of the techniques used to establish the viscous flow properties discussed in the previous chapter. One of the common methods, Poiseuille (laminar) flow in cylindrical tubes is also important from a technological standpoint, as polymer melts and solutions are often transported and processed in this fashion. Because of their tremendous viscosities, laminar flow is far more prevalent with polymeric fluids than with nonpolymer fluids; in fact, turbulence is normally encountered only with dilute (less than a few weight percent, or so) polymer solutions.

16.2 VISCOUS ENERGY DISSIPATION

Regardless of the viscometric technique, determination of the true, isothermal flow properties at high shear rates can be very difficult because of the high rates of viscous energy dissipation, which makes it hard to maintain isothermal conditions. The *rate of viscous energy dissipation per unit volume*, $\dot{E}$, is

$$\dot{E} = \tau\left(\frac{\text{dynes}}{\text{cm}^2}\right)\dot{\gamma}\left(\frac{1}{\text{sec}}\right) \times 1\left(\frac{\text{cm}}{\text{cm}}\right) = \tau\dot{\gamma}\left(\frac{\text{dyne-cm}}{\text{cm}^3\text{-sec}}\right) = \tau\dot{\gamma}\left(\frac{\text{ergs}}{\text{cm}^3\text{-sec}}\right).$$

(16-1)

Example 1. Obtain an expression for the adiabatic rate of temperature rise in a polymer sample subjected to a shear stress τ and shear rate $\dot{\gamma}$.

Solution.

$$\frac{dT}{dt} = \tau\dot{\gamma}\left(\frac{\text{ergs}}{\text{cm}^3\text{-sec}}\right) \times \frac{\rho\,(\text{grams/cm}^3)}{C_p(\text{ergs/gram-}°\text{C})} = \frac{\tau\dot{\gamma}\rho}{C_p}\left(\frac{°\text{C}}{\text{sec}}\right) \quad (16\text{-}2)$$

Not only is this an important consideration in viscometry, where great care must be taken in the design of viscometers to permit adequate temperature regulation, but it also must be taken into account in the design of processing systems. In the steady-state operation of extruders, for example, virtually all the energy required to melt and maintain the polymer in the molten state is supplied by the mechanical drive. Here, however, we will limit our considerations to isothermal flow situations.

16.3 POISEUILLE FLOW

Axial, laminar (Poiseuille) flow in a tube of cylindrical cross section is an example of the viscometric flow field discussed in the previous chapter. Here, the geometry of the situation dictates the use of cylindrical coordinates. Fluid motion is in the x ("1") direction along the tube axis, the velocity gradient is everywhere directed in the outward radial r (or 2) direction, and the mutually perpendicular or neutral direction is the θ (or 3) coordinate. If the fluid pressure is a function of distance along the tube (x coordinate) only, by equating the shear force on the surface of a cylindrical fluid element of radius r and length dx to the axial forces on it arising from the differential pressure drop $-dP$ across its ends (Figure 16.1) (they must balance in an equilibrium flow situation), we find that

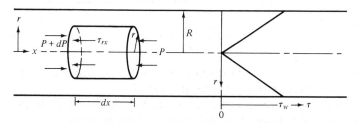

Figure 16.1. Force balance on an element in a cylindrical tube.

$$2\pi r\, dx\, \tau_{rx} \;=\; -\pi r^2\, dP. \qquad (16\text{-}3)$$

surface shear   net pressure
force   force

Thus, the dependence of shear stress on radius (denoted $\tau(r)$) is

$$\tau_{rx}(r) \;=\; -\frac{r}{2}\left(\frac{dP}{dx}\right). \qquad (16\text{-}4)$$

For a tube with inner radius R (dropping the subscripts on shear stress)

$$\tau_w \;=\; -\frac{R}{2}\left(\frac{dP}{dx}\right) \qquad (16\text{-}5)$$

where τ_w is the shear stress at the tube wall, and

$$\tau(r) \;=\; \left(\frac{r}{R}\right)\tau_w. \qquad (16\text{-}6)$$

So, the shear stress varies linearly with radius from zero at the tube center to a maximum of $\tau_w = -(R/2)(dP/dx)$ at the tube wall. *Note that this result does not depend in any way on the fluid properties.*

Since the axial fluid velocity u is a function of radial position only (denoted $u(r)$), we may write

$$du(r) \;=\; \left(\frac{du}{dr}\right)dr. \qquad (16\text{-}7)$$

By using the boundary conditions $u(R) = 0$ (i.e., the fluid sticks to the wall), and u at radius $r = u(r)$ and integrating,

$$u(r) \;=\; \int_{o}^{u(r)} du(r) \;=\; \int_{R}^{r}\left(\frac{du}{dr}\right)dr. \qquad (16\text{-}8)$$

Since the velocity gradient (du/dr) is simply the shear rate $\dot{\gamma}$,

$$u(r) \;=\; \int_{R}^{r} \dot{\gamma}(\tau)\, dr. \qquad (16\text{-}9)$$

where τ, in turn, is a function of r (16-4). The function $\dot{\gamma}(\tau)$ is, of course, the flow curve for the material.

Thus, for a given pressure gradient, the velocity profile in the tube may always be calculated from the flow curve by

1. choosing an r and calculating τ from (16-4),

2. obtaining $\dot{\gamma}$ at the τ above from the flow curve,

3. plotting $\dot{\gamma}$ versus r (or otherwise numerically representing the relation), and

4. integrating from R to r, which gives u at r.

If an analytic representation of the flow curve is available, (16-9) may be integrated directly (provided the representation is simple enough). For example (verification will be left as an exercise for the reader), for a power-law fluid $\dot{\gamma}(\tau) = (\tau/K)^{1/n}$, and

$$u(r) = \frac{[(-dP/dx)/2K]^{1/n}}{(1/n) + 1} [R^{(1/n)+1} - r^{(1/n)+1}]. \quad (16\text{-}10)$$

Equation (16-10) reduces to the usual Newtonian parabolic profile for $n = 1$.

Looking at a differential ring of the tube cross section, with thickness dr, located at radius r where the velocity is $u(r)$ (Figure 16.2), the differential volumetric flow rate is

$$dQ = \underbrace{u(r)}_{\substack{\text{local} \\ \text{velocity}}} \quad \underbrace{2\pi r\,dr.}_{\substack{\text{area of} \\ \text{differential ring}}} \quad (16\text{-}11)$$

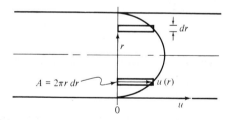

Figure 16.2.   Determination of volumetric flow rate.

By integrating over the tube cross section,

$$Q = 2\pi \int_{o}^{R} u(r)\,r\,dr \left(\frac{cm^3}{sec}\right). \quad (16\text{-}12)$$

Thus, knowing the velocity profile allows calculation of the volumetric throughput, Q. For a power-law fluid,

$$Q = \left[\frac{-(dP/dx)}{2K}\right]^{1/n}\left(\frac{\pi}{(1/n) + 3}\right)R^{(1/n)+3}. \qquad (16\text{-}13)$$

The *average velocity* in the tube is defined by

$$V \equiv Q/\pi R^2. \qquad (16\text{-}14)$$

Now let's take a look at velocity profiles in the tube for power-law fluids, plotted in dimensionless form in Figure 16.3 for

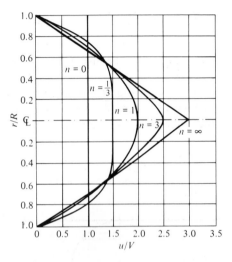

Figure 16.3. Velocity profiles for the laminar flow of power-law fluids in tubes.

several values of n. The greater the degree of pseudoplasticity (i.e., the lower n), the flatter the profile becomes. For the Newtonian fluid, $n = 1$, the profile is parabolic.

16.4 DETERMINATION OF FLOW CURVES

The formulas just developed allow the relation of pressure gradient, velocity profile and volumetric flow rate *as long as the shear stress-shear rate relation* (flow curve) *for the fluid is known*. The problem in viscometry is just the reverse; i.e., how is the flow curve obtained from pressure drop-volumetric flow rate measurements in a cylindrical tube? *If* the mathematical form of the flow curve (i.e., a particular constitutive equation) *is assumed a priori,* the integrated equations as developed above may be used to establish the parameters in the constitutive relation. For example, if it is assumed that the power law represents the flow curve of a fluid under investigation, two readings of dP/dx versus Q will allow calculation of K and n from (16-13). However, in the general case, the form of the constitutive equation is not known a priori and must be established by viscometry. This may be done, first by integrating (16-12) by parts

$$Q = 2\pi \left[\frac{r^2 u(r)}{2} \bigg|_o^R - \int_o^R \frac{r^2}{2} \, du(r) \right]. \qquad (16\text{-}15)$$

Fortunately, the first term in the brackets of (16-15) is zero because, at the upper limit, $u(R) = 0$, and, at the lower limit, $r = 0$. Therefore,

$$Q = -\pi \int_o^R r^2 \, du(r) = -\pi \int_o^R r^2 \left(\frac{du(r)}{dr} \right) dr = -\pi \int_o^R r^2 \dot{\gamma}(r) \, dr. \qquad (16\text{-}16)$$

By using (16-6) to change the independent variable from r to τ (and recalling that at $r = 0$, $\tau = 0$ and at $r = R$, $\tau = \tau_w$),

$$\frac{\tau_w^3 Q}{\pi R^3} = - \int_o^{\tau_w} \dot{\gamma}(\tau) \tau^2 \, d\tau. \qquad (16\text{-}17)$$

Applying Leibnitz' rule to differentiate both sides of (16-17) with respect to τ_w gives

$$\frac{1}{\pi R^3} \left[\tau_w^3 \frac{dQ}{d\tau_w} + 3\tau_w^2 Q \right] = -\dot{\gamma}_w \tau_w^2 \qquad (16\text{-}18)$$

or,

$$-\dot{\gamma}_w = \frac{1}{\pi R^3} \left[\tau_w \frac{dQ}{d\tau_w} + 3Q \right]. \qquad (16\text{-}19)$$

Since $\tau_w = -(R/2)(dP/dx)$, $(\tau_w/d\tau_w) = (dP/dx)/d(dP/dx)$, and

$$-\dot{\gamma}_w = \frac{1}{\pi R^3} \left[3Q + \left(\frac{dP}{dx} \right) \frac{dQ}{d(dP/dx)} \right]. \qquad (16\text{-}20)$$

Also, since $d \ln \xi = d\xi/\xi$ (ξ = any variable),

$$-\dot{\gamma}_w = \frac{4Q}{\pi R^3} \left[\frac{3}{4} + \frac{1}{4} \frac{d \ln Q}{d \ln (dP/dx)} \right]. \qquad (16\text{-}21)$$

Equations (16-20) and (16-21) are equivalent forms of the *Rabinowitsch equation*, allowing calculation of the shear rate at the wall of the tube from experimental data. Note the necessity in (16-21) of obtaining the slope of a log Q versus log (dP/dx) plot (or otherwise differentiating the data) at each value of Q to obtain $\dot{\gamma}_w$ for that Q. Values of $\dot{\gamma}_w$ are then combined with values of τ_w calculated from (16-5) at the same Q to give the flow curve.

The quantity $(4Q/\pi R^3) = (8V/D)$ is known as the *apparent shear rate*. *Only for Newtonian fluids, where* $d \ln Q/[d \ln (dP/dx)] = 1$, *is it the true shear rate at the wall of the tube*. Neglect of the *Rabinowitsch correction* (the term in square brackets in (16-21)) can lead to appreciable error. To illustrate, for a power-law fluid, $d \ln Q/[d \ln (dP/dx)] = 1/n$, a constant. The Rabinowitsch equation then becomes

$$-\dot{\gamma}_w = \frac{4Q}{\pi R^3} \left[\frac{3n + 1}{4n} \right] \quad \text{(Power-law fluids)}. \qquad (16\text{-}22)$$

For an n of $1/3$, a typical value for a polymer melt or solution, the correction term is 1.5, so the apparent shear rate is 50 percent lower than the true value at the tube wall. The Rabinowitsch correction accounts for the fact that the velocity gradient (shear rate) is greater at the tube wall for pseudoplastic fluids than for a Newtonian fluid at a given Q, as illustrated in Figure 16.3.

When comparing viscometric data from tubes with that obtained from other types of viscometers and when using tube flow data to predict performance in other geometries, the use of the Rabinowitsch correction is essential to give the true shear stress-shear rate relation. If all you're concerned with is flow in cylindrical tubes, however, it's not necessary. A look at (16-17) shows that, for the flow of a given fluid in cylindrical tubes, τ_w is a unique function of the apparent shear rate $4Q/\pi R^3 = 8V/D$, because the value of the integral depends only on τ_w regardless of the nature of the $\dot{\gamma}(\tau)$ relation (flow curve) of the fluid. Thus, experimental determinations of τ_w versus $4Q/\pi R^3 = 8V/D$ obtained from one cylindrical tube are applicable for scaleup purposes, etc., to any other cylindrical tube through which the same fluid is flowing (see Ref. (1), Example 3-2).

16.5 ENTRANCE CORRECTIONS

A capillary viscometer is a device in which the fluid under investigation is forced from a reservoir into a cylindrical capillary tube. They are operated with either flow rate or pressure as the independent variable. In the former, the fluid is usually driven by a piston advancing through the reservoir at a known constant rate with the force on the piston recorded; in the latter, regulated gas pressure drives the fluid and the volumetric flow rate is measured. Unfortunately, calculating the pressure gradient from the data so obtained is not always a simple matter of setting $(dP/dx) = \Delta P/L$, where ΔP is the measured overall pressure drop across the capillary of length L. The gradient does not become constant until equilibrium flow is reached, some distance downstream from the entrance of the capillary (as illustrated in Figure 16.4), and approximating the true equilibrium gradient (dP/dx) with the measured $\Delta P/L$ can cause considerable error. Higher-than-equilibrium pressure losses are observed in the entry region because (1) kinetic energy must be added to the fluid as it is accelerated from low velocities in the reservoir to higher velocities in the capillary, (2) rearrangement of the velocity profile dissipates additional energy, and (3) a viscoelastic fluid *stores* some energy elastically when going from the low-stress reservoir to the high-stress capillary (think of stretching a rubber band). Recovery of this stored energy at the tube exit leads to such things as die swell, melt fracture, and other anomalies observed with the flow of viscoelastic fluids.

Bagley (2) has developed a method for getting the true pressure gradient from capillary viscometer data. As seen in Figure 16.4,

$$\left(\frac{dP}{dx}\right) = \frac{\Delta P}{(L + L_e)} = \frac{\Delta P}{(L + eR)} \qquad (16\text{-}23)$$

where L_e is a fictitious entrance length. L_e is commonly expressed as the product of e, the *entrance correction* and the tube

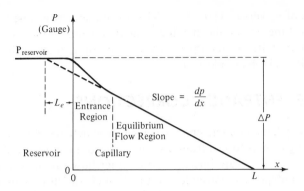

Figure 16.4.   Pressure profiles in a capillary viscometer.

radius R. Equation (16-5) can be rewritten, by dividing numerator and denominator by R and recalling that τ_w is a function of $4Q/\pi R^3$ *only*, as

$$\tau_w = \frac{R\Delta P}{2(L + eR)} = \frac{\Delta P}{2((L/R) + e)} = f\left(\frac{4Q}{\pi R^3}\right) \quad (16\text{-}24)$$

where f denotes 'function of.' After rearranging,

$$\Delta P = 2f\left(\frac{4Q}{\pi R^3}\right)\left(\frac{L}{R} + e\right). \quad (16\text{-}25)$$

Thus, if a series of experiments is run in which the capillary L/R ratio is varied while holding the apparent shear rate $(4Q/\pi R^3)$ constant, a *Bagley plot* of ΔP versus L/R should give a series of straight lines, one for each constant value of $(4Q/\pi R^3)$ investigated, with the $\Delta P = 0$ intercept the entrance correction, e (Figure 16.5). The various intercepts at constant apparent shear rates give e as a function of τ_w or $\dot\gamma_w$. The need to use several capillaries greatly increases the work, but, for melts, it appears that $(L/R) \gg e$ (i.e., the entrance correction becomes negligible) only for $(L/R) > 100$ or so. For solutions, at high flow rates, the correction may be significant at much higher L/R's. The reverse of this procedure may be important in design situations; i.e., the entrance correction must be known as a function τ_w and included in the overall pressure drop calculations. In extruder dies and spinerettes for fiber spinning, the L/R's rarely exceed 10, and the entrance

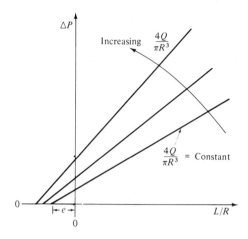

Figure 16.5.   Bagley Plot to determine entrance corrections.

loss may be the major contribution to the overall pressure drop. Carley (*3*) provides examples of design calculations involving these concepts.

Once the entrance correction has been established, it can be used to calculate the true equilibrium gradient through (16-23), which in turn is used in (16-5) and (16-21) to obtain the flow curve, relating the shear stress to shear rate, both at the tube wall. In a few instances, other minor corrections have been applied—e.g., considering the pressure drop in the viscometer reservoir (*4*).

16.6 THE COUETTE VISCOMETER

Another common device for measuring viscous properties is the cup-and-bob (or Couette) viscometer, a diagram of which is shown in Figure 16.6. The fluid is confined in the gap between two concentric cylinders, one of which moves relative to the other at a known angular velocity while the torque on one is measured. This is another classic example of a viscometric flow, the 1 coordinate being the tangential or θ direction and the 2 coordinate being the radial direction.

Example 2. a. Neglecting end effects, determine the shear stress as function of radius in terms of the measured torque,

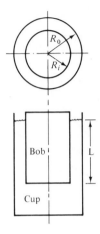

Figure 16.6.   Schematic of Couette (cup-and-bob) viscometer.

M, on the stationary inner cylinder (bob) and the geometry of the apparatus as the outer cylinder (cup) is rotated with an angular velocity ω (radians/sec).

b. The problem of determining the shear rate as a function of radius is not an easy one. Bird, Steward and Lightfoot (5) do it by *assuming* a Newtonian relation between shear stress and shear rate. In general, however, the nature of the flow curve is not known a priori and must be determined by viscometry. One means of doing this is to make the gap between the cylinders $(R_o - R_i)$ very small compared to the radius of either cylinder. By letting $(R_o - R_i) = d$ and $R_o \simeq R_i = R$, obtain the expression for shear rate in terms of ω and the geometry.

Solution. a. In a rotating system at equilibrium, Σ torques = 0, or there would be angular acceleration. Consider a ring of fluid with inner radius R_i and outer radius r: $M(r) = M(R_i)$

$$\underbrace{\underbrace{(2\pi r L)(\tau(r))}_{\substack{\text{surface} \\ \text{area}}} \underbrace{(r)}_{\substack{\text{moment} \\ \text{arm}}}}_{\text{force}} = M(R_i)$$

$$\tau(r) = \frac{M}{2\pi r^2 L} \qquad (16\text{-}26)$$

b. This situation approximates the case of two flat plates separated by a distance d, one sliding past the other with a linear velocity equal to the tangential velocity $R\omega$

$$\dot{\gamma} = \frac{R\omega}{d} \quad \text{(for } d \ll R_i \text{ only).} \qquad (16\text{-}27)$$

Where the geometric approximations in example 2b above are not applicable, Kreiger and Maron (6) have presented an analysis similar to the Rabinowitsch development for flow in tubes. It involves differentiation of the M versus ω data but, unfortunately, is in the form of an infinite series. If $R_o/R_i < 1.2$, a closed-form approximation is available, however.

Obviously, the analysis above is not valid in the area beneath the bob at the bottom of the viscometer. This is normally taken into account by calibrating the viscometer with a Newtonian fluid of known viscosity, establishing an effective length for the instrument or, preferably, making measurements with two fluid depths, and using the *differences* between the torques and heights in (16-26) above. In some Couette viscometers, the bottom of the bob is concave, trapping a bubble of air which does not add appreciably to the measured torque.

16.7 CONE-AND-PLATE VISCOMETER

Still another popular type of viscometer is the cone-and-plate, in which the sample is sheared between a flat plate and a broad cone whose apex contacts the plate (Figure 16.7). For

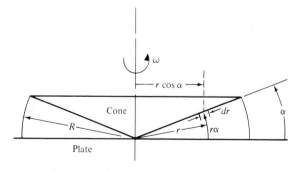

Figure 16.7.   Schematic of cone-and-plate viscometer.

small cone-plate angles, α, this approximates a viscometric flow
with (in spherical coordinates) flow in the tangential θ or 1
direction and the gradient in the azimuthal ϕ or 2 direction.
Here, the radial direction is the neutral or 3 coordinate. (It
may be shown that true viscometric flow of this type is incon-
sistent with the equations of motion if the inertial terms are
included. There must, therefore, be radial and azimuthal
velocity components. These are minimized in practice by keeping α quite small—often less than 1°.) The great advantage
of this type of device is that (for small α) the shear rate, and
hence the shear stress, is constant throughout the gap.

Example 3. Obtain the expressions for shear rate and shear
stress in a cone-and-plate viscometer in terms of the rate of
cone rotation ω, the measured torque, M, and the geometry.

Solution (see Figure 16.7). The tangential velocity v of a
point on the cone relative to the plate is $v = \omega r \cos \alpha$. Fluid
is sheared between that point and the plate over a distance
$d = \alpha r$

$$\dot{\gamma} = \frac{v}{d} = \frac{\omega r \cos \alpha}{\alpha r} \underset{\alpha}{\overset{small}{=\!=}} \frac{\omega}{\alpha} \quad (independent \ of \ r) \qquad (16\text{-}28)$$

$$\text{torque} = \underbrace{(\text{shear stress})(\text{area})}_{\text{force}}(\text{moment arm})$$

$$dM = (\tau)(2\pi r dr)(r \cos \alpha)$$

integrating over the cone face

$$M = 2\pi\tau \cos \alpha \int_0^{R/\cos \alpha} r^2 \, dr = \frac{2\pi\tau R^3}{3\cos^2 \alpha} \underset{\alpha}{\overset{small}{=\!=}} \frac{2\pi\tau R^3}{3}$$

$$\tau = \frac{3M}{2\pi R^3} \qquad (16\text{-}29)$$

Note: here, τ is constant and can be taken outside the integral,
because $\tau = f(\dot{\gamma})$, and $\dot{\gamma}$ is constant, as shown by
(16-28).

Cone-and-plate viscometers of the type shown here are usually limited to fairly low shear rates. At higher shear rates,
solutions tend to be flung from the gap by centrifugal force,

and melts tend to ball up (like rubbing a finger over dry rubber cement). These problems can be overcome by enclosing the fluid around a biconical rotor, giving, in effect, two cone-and-plate viscometers back-to-back (7). The flow curves in chapter 15 were obtained with such a device.

16.8 TURBULENT FLOW

On occasions, turbulent flow is encountered with dilute polymer solutions. As with the turbulent flow of Newtonian fluids, pressure drops are conveniently handled in terms of the Fanning friction factor

$$f = \frac{Rg_c}{\rho V^2}\left(\frac{dP}{dx}\right). \tag{16-30}$$

(The dimensional constant $g_c = 32.2$ ft-lb$_m$/lb$_f$-sec^2 is included here because these equations are most often used with the English engineering system of units.) Since $\tau_w = (R/2)(dP/dx)$ (the minus sign has been dropped, it being understood that flow is in the direction of decreasing pressure),

$$f = \frac{2\tau_w g_c}{\rho V^2}. \tag{16-31}$$

Equations (16-30) and (16-31) are in no way dependent on the nature of the fluid. If, as suggested by Metzner and Reed (8), the analogy between the *laminar flow* of non-Newtonian and Newtonian fluids is to be preserved; then the usual relation

$$f = \frac{16}{Re} \tag{16-32}$$

where Re is the Reynolds number, is applicable to both. By combining (16-31) and (16-32), the *generalized Reynolds number*—for any fluid—becomes

$$Re = \frac{8\rho V^2}{\tau_w g_c}. \tag{16-33}$$

For a power-law fluid, $\tau g_c = K(\dot{\gamma})^n$, (16-5), (16-13) and (16-14) give

$$Re = \frac{8 D^n V^{2-n} \rho}{K(6 + 2/n)^n}. \qquad (16\text{-}34)$$

Actually, Metzner and Reed used an alternate formulation. Since the shear stress at the tube wall is a unique function of the apparent shear rate for laminar flow in tubes, the power law may be written at the tube wall with the aid of (16-22) as

$$\tau_w g_c = K(\dot{\gamma}_w)^n = K\left(\frac{4Q}{\pi R^3}\left[\frac{3n + 1}{4n}\right]\right)^n = K'\left(\frac{4Q}{\pi R^3}\right)^{n'} \quad (16\text{-}35)$$

where

$$n' = n \quad \text{and} \quad K' = K\left[\frac{3n + 1}{4n}\right]^n. \qquad (16\text{-}36)$$

With these variables, the generalized Reynolds number becomes

$$Re = \frac{D^{n'} V^{2-n'} \rho}{K' 8^{n'-1}}. \qquad (16\text{-}37)$$

Both (16-34) and (16-37) reduce to the usual Newtonian relation, $Re = DV\rho/\eta$, with $\eta = K = K'$, when $n = n' = 1$.

Equations (16-30), (16-31), (16-32) and (16-33) *by definition* should describe the behavior of all fluids in *equilibrium* laminar viscous flow. They do, *provided that the true equilibrium pressure gradient* is used in (16-30). Approximation of the gradient by $\Delta P/L$ can cause considerable error. There is also much older data to suggest that the usual friction-factor versus Reynolds number charts are applicable to turbulent non-Newtonian flow, with the variables defined as above. The laminar-turbulent transition occurs at about $Re = 2100$, as with Newtonians.

16.9 DRAG REDUCTION (9, 10)

Considerable newer research has revealed a startling phenomenon known as *drag reduction* in the turbulent flow of non-Newtonians. Drag reduction is generally observed with dilute polymer solutions—so dilute, in fact, that their viscosities are often not noticeably different from that of the solvent alone. As shown in Figure 16.8 (9), the data are usually pre-

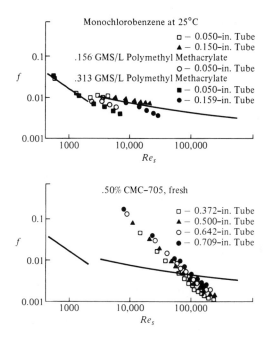

Figure 16.8.   Examples of turbulent drag reduction (9). (Copyright 1967 by the American chemical society. Reprinted by permission of the copyright owner.)

sented in the form of friction factor (16-30) versus the *solvent* Reynolds number, $Re_s = DV\rho/\eta_{solvent}$. The minute quantities of polymeric solute can cut the friction factor significantly— by a factor of two or more—at higher Reynolds numbers. This phenomenon has important practical applications. Pressure drops and, therefore, pumping costs can be reduced for a given flow rate, or the capacity of a pumping system can be increased by the addition of a drag-reducing solute.

The causes of turbulent drag reduction are not yet clear. In some cases, the additive seems to stabilize laminar flow, with its lower f, to much higher Re than usual. In others, the friction factor drops away from the usual curve at higher Reynolds numbers, and, in still others, it just cuts across the normal curve as the Reynolds number increases. It has been suggested that the phenomenon may be due to thickening of the laminar sublayer. Another possible cause is the viscoelasticity of polymer solutions. Most of the energy dissipation in turbulent flow normally occurs in small, high-frequency eddies. It has been suggested that, at these frequencies (the reciprocal of which are smaller than the material's relaxation time, see chapter 18), the material responds elastically (like bouncing the silly putty rather than letting it flow down the wall), passing stored elastic energy from eddy to eddy rather than dissipating it. In any case, the sad fact remains that, at present, there are no concrete methods for predicting which polymeric solutes will produce drag reduction in a particular solvent or over what range of concentrations it will be observed and what its magnitude will be.

REFERENCES

1. McKelvey, J. M. *Polymer Processing*, ch. 3. John Wiley & Sons, Inc., New York, 1962.

2. Bagley, E. B. End corrections in the capillary flow of polyethylene. *J. Appl. Phys.*, **28,** no. 624, 1957.

3. Carley, J. F. Problems of flow in extrusion dies, *Soc. Plast. Eng. J.*, **19,** no. 12, p. 1263, 1963.

4. Van Wazer, et. al. *Viscosity and Flow Measurement—A Laboratory Handbook of Rheology.* John Wiley & Sons, Inc., New York, 1963.

5. Bird, R. B., W. E. Stewart and E. N. Lightfoot. *Transport Phenomena*, p. 94 (example 3.5-1). John Wiley & Sons, Inc., New York, 1960.

6. Kreiger, I. M. and S. H. Maron. Direct determination of the flow curves of non-newtonian fluids III. *J. Appl. Phys.*, **25,** no. 72, 1954.

7. Best, D. M. and S. L. Rosen. A simple, versatile and inexpensive rheometer for polymer melts. *Polymer Engineering and Science*, **8,** no. 2, p. 116, 1968.

8. Metzner, A. B. and J. C. Reed. Flow of non-Newtonian fluids—correlation of the laminar, transition and turbulent-flow regions. *Amer. Inst. Chem. Eng. J.*, **1,** no. 434, 1955.

9. Hershey, H. C. and J. L. Zakin. Existence of two types of drag reduction in pipe flow of dilute polymer solutions. *I&EC Fundamentals*, **6,** no. 3, p. 381, 1967.

10. Patterson, G. K., J. L. Zakin and J. M. Rodriguez. Drag reduction. *Ind. and Eng. Chem.*, **61,** no. 1, p. 22, 1969.

17
Introduction to Continuum Mechanics

17.1 THREE-DIMENSIONAL STRESS AND STRAIN (1, 2)

The stress existing at any point in a material may always be resolved into components acting on the faces of a differential element in three arbitrary coordinate directions (Figure 17.1). The stress components acting on the faces of the element are of two types, *normal stresses* (forces *normal* to the surface per unit surface area) and shear stresses (forces *parallel* to the surface per unit surface area). The first subscript conventionally identifies the direction perpendicular to the surface in question, and the second identifies the direction of the force itself. In general, there are three *normal* stresses, τ_{11}, τ_{22} and τ_{33}, and six *shear* stresses, τ_{12}, τ_{13}, τ_{21}, τ_{23}, τ_{31} and τ_{32}. These nine quantities, necessary to specify the state of stress at a point completely, are the components of the *stress tensor* τ. They are conveniently written in matrix form:

$$\tau = \begin{vmatrix} \tau_{11} & \tau_{12} & \tau_{13} \\ \tau_{21} & \tau_{22} & \tau_{23} \\ \tau_{31} & \tau_{32} & \tau_{33} \end{vmatrix} \tag{17-1}$$

The stress tensor is usually broken into an isotropic or hydrostatic pressure and a *deviatoric stress tensor*:

$$\tau = - \begin{vmatrix} P & 0 & 0 \\ 0 & P & 0 \\ 0 & 0 & P \end{vmatrix} + \begin{vmatrix} p_{11} & \tau_{12} & \tau_{13} \\ \tau_{21} & p_{22} & \tau_{23} \\ \tau_{31} & \tau_{32} & p_{33} \end{vmatrix} \tag{17-2}$$

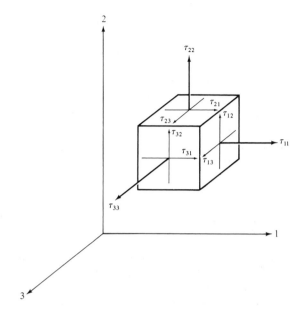

Figure 17.1. Stress components acting on an element of material. The equal-and-opposite stresses (necessary for equilibrium) on the hidden faces are not shown.

where P, the hydrostatic pressure is rather arbitrarily defined as

$$P = -\frac{(\tau_{11} + \tau_{22} + \tau_{33})}{3} \qquad (17\text{-}3)$$

and $p_{ii} = \tau_{ii} + P$ are the *deviatoric normal stresses*.

Caution: (a) The negative signs in (17-2) and (17-3) arise from the historical conventions of treating both hydrostatic pressure (an inward force on the element) and *tensile* stress (an outward force on the element) as positive. This convention is not adhered to universally, however.

Caution: (b) There are about as many different forms of notation as there have been books and papers written in this area. Be sure to understand the notation before plowing through any of the literature—that's half the battle.

17.2 GENERAL CONSTITUTIVE RELATIONS

The three-dimensional rate of strain at the point is expressed by a rate-of-strain tensor analogous to (17-1). Given a general *constitutive relation* between the stress and rate-of-strain tensors for a particular material, the mechanical response of that material can be described completely (in principle) under any circumstances. The goal of continuum mechanics is to develop such general constitutive relations and use them for predicting material response in the widest variety of situations. To complete the picture ultimately, it is the goal of molecular mechanics to predict these constitutive relations from the molecular structure of the material. Unfortunately, from the standpoint of engineering application, the attainment of these goals is a long way off.

17.3 VISCOMETRIC FLOWS (1)

For *equilibrium viscometric* flows of *incompressible* fluids (the latter is a reasonable assumption for polymer melts and solutions in most engineering circumstances), the situation is brighter. With the coordinate directions assigned as described for viscometric flows in the last chapter, the shear stresses τ_{12} and τ_{21} are equal (or there would be a rotational flow component), and $\tau_{13} = \tau_{31} = 0$. There are two independent differences of the deviatoric normal stresses, commonly defined as

$$\sigma_1 = p_{11} - p_{22} \tag{17-4a}$$

$$\sigma_2 = p_{22} - p_{33} \tag{17-4b}$$

where σ_1 and σ_2 are the so-called first and second normal stress differences. Furthermore, there is only one nonzero component of the rate-of-strain tensor, $\dot{\gamma}_{12} = \dot{\gamma}$. Thus, a complete description of an incompressible material in an equilibrium viscometric deformation requires only a knowledge of the three functions

$$\tau = \tau(\dot{\gamma}) \quad \text{(the flow curve)} \tag{17-5a}$$

$$\sigma_1 = \sigma_1(\dot{\gamma}) \tag{17-5b}$$

$$\sigma_2 = \sigma_2(\dot{\gamma}) \tag{17-5c}$$

i.e., the dependence of shear stress and first and second normal stress differences on the shear rate. Techniques for measuring functions (17-5b) and (17-5c) have been comprehensively reviewed (*1*).

In nonequilibrium deformations, the response of the material depends not only on its present rate of strain but on its complete strain history. Elaborate and very general constitutive relations have been developed to handle these situations, but they have achieved little engineering application because of their unavoidable complexity.

Example 1. Consider the classical example of a nonviscometric flow, *simple elongation*. This situation arises, for example, if a weight is suspended from a rod of material. Note that here the velocity gradient *is not* perpendicular to the direction of fluid motion. Discuss the nature of the stress and rate of strain tensors.

Solution. Neglecting such things as surface tension forces on the sides of the rod, the deviatoric stress tensor has only one component, τ_{11}, where the subscript 1 represents the axial direction. This will obviously produce a tensile elongation, so there will be a rate of tensile elongation component of the rate-of-strain tensor $\dot{\epsilon}_{11}$. For an *incompressible* material (Poisson's ratio = 0.5), the lateral dimension of the rod will contract to maintain the volume constant as the rod is extended, and the lateral *contractile* strains will be one-half the axial extension strain. Thus $\dot{\epsilon}_{11} = -2\dot{\epsilon}_{22} = -2\dot{\epsilon}_{33}$. There is no shearing, so all the shear strains are zero.

Therefore, to describe this sort of deformation, a new material function η_e, the elongational or Trouton viscosity is needed to relate tensile stress to the rate of tensile strain; i.e., $\tau_{11} = \eta_e(\dot{\epsilon}_{11})\dot{\epsilon}_{11}$, or equivalently the function $\tau_{11} = \tau_{11}(\dot{\epsilon}_{11})$. For *incompressible Newtonian fluids*, it may be shown that $\eta_e = 3\eta$.

17.4 INTERPRETATION OF THE NORMAL STRESSES

The presence of deviatoric normal stress components in viscometric flow implies fluid elasticity, and they are physically real quantities. The well-known *Weissenberg effect* is a good example. When a vertical rotating rod is immersed in a container of inelastic fluid, the fluid is flung outward by centrifugal force, and its surface assumes a parabolic profile with the lowest point at the rod. A viscoelastic fluid, on the other hand, will actually climb up the rod. This is a viscometric Couette flow, as described in the previous chapter. The behavior of the viscoelastic fluid can be rationalized by imagining it to consist of rubber bands, stretched by the rotation along a fluid streamline in the tangential (θ or 1) direction at constant radius. The rubber bands are obviously in tension, and hence $p_{11} = p_{\theta\theta}$ is positive (a hoop stress, as in the bands of a barrel). This positive hoop stress places the cylinder of material within under a compressive stress in the radial direction (think of what wrapping a rubber band around your finger does); i.e., a negative $p_{22} = p_{rr}$. Since the inwardly compressed material is prevented from moving inward by the presence of the rod, it has nowhere to go but up the rod under the influence of a compressive axial stress; i.e., a negative $p_{33} = p_{zz}$. This does not mean to imply that centrifugal force is no longer acting but merely that it is overwhelmed by the elastic normal stresses.

Similar reasoning leads to the conclusion that in a cone-and-plate viscometer, or in a torsional flow, with a disk rotating parallel to a flat plate, there will be forces tending to push the cone and plate or the disk and plate apart. Indeed, there are, and the measurement of these forces provides a means of determining the functions $\sigma_1(\dot{\gamma})$ and $\sigma_2(\dot{\gamma})$ (1). Also, for the case of torsional flow, Maxwell and Scalora (3) showed that, if a hole is drilled through the plate along the axis of rotation, a screwless extruder is obtained, as the axial normal stress pushes a polymer melt through the hole, (Figure 17.2). Unlike a screw extruder, however, this gizmo only works with viscoelastic fluids.

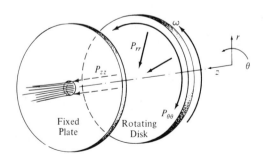

Figure 17.2.   Normal stresses in the elastic melt extruder (3).

REFERENCES

1. Coleman, B. D., H. Markovitz and W. Noll. *Viscometric flow of non-newtonian fluids.* Springer-Verlag Inc., New York, 1966.

2. Middleman, S. *The flow of high polymers.* John Wiley & Sons, Inc., New York, 1968.

3. Maxwell, B. and A. J. Scalora. The elastic melt extruder-works without screw. *Mod. Plast.*, **37**, no. 107, 1959.

18
Linear Viscoelasticity

18.1 INTRODUCTION

Engineers have traditionally been concerned with two separate and distinct classes of materials—the viscous fluid and the elastic solid. Elaborate design procedures have been based on these concepts and have worked pretty well because most traditional materials (water, motor oil, steel, concrete), at least to a good approximation, fall in one of these categories. The realization has grown, however, that these categories represent only the extremes of a broad spectrum of material response, and polymers fall somewhere in between, giving rise to some of the unusual properties of melts and solutions described previously. Other examples are important in the structural applications of polymers. In a common engineering stress-strain test, a sample is strained at an approximately constant rate, and the stress is measured as a function of strain. With traditional solids, the stress-strain curve is pretty much independent of the rate at which the material is strained. The stress-strain properties of many polymers are markedly *rate dependent*, however. Similarly, polymers often exhibit pronounced creep and stress relaxation (to be defined shortly). While other materials also exhibit such behavior (metals near their melting points, for example), at normal temperatures it is negligible and is not usually included in design calculations. If the time-dependent behavior of polymers is ignored, the results can sometimes be disastrous.

204

18.2 MECHANICAL MODELS FOR LINEAR VISCOELASTIC RESPONSE

As an aid in visualizing viscoelastic response, we introduce two *linear* mechanical models to represent the extremes of the mechanical response spectrum. Figure 18.1a will represent a

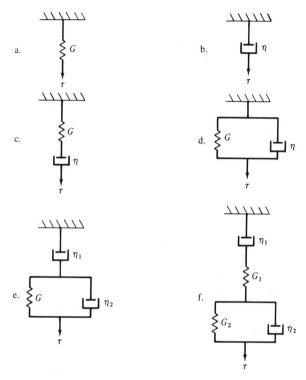

Figure 18.1.   Linear viscoelastic models. (a) linear elastic, (b) linear viscous, (c) Maxwell element, (d) Voigt-Kelvin element, (e) three parameter, (f) four parameter.

linear elastic or Hookean *solid* whose constitutive equation (relation of stress to strain and time) is simply $\tau = G\gamma$, where G is a (constant) shear modulus. Similarly, a *linear viscous* or Newtonian *fluid* is represented with a dashpot (some sort of piston moving in a cylinder of Newtonian fluid) whose constitutive equation is $\tau = \eta\dot{\gamma}$, where η is a (constant) viscosity. Although the models developed here are to be visualized in tension, the notation used is for pure shear (viscometric) deformation. The equations are equally applicable to tensile deformation by replacing the shear stress τ with the tensile stress σ, the shear strain γ with the tensile strain ϵ, Hooke's modulus G with Young's (tensile) modulus E, and the Newtonian (shear) viscosity η with Trouton's (tensile) viscosity η_e.

Some authorities object strongly to the use of mechanical models to represent materials. They point out that materials are not made up of springs and dashpots. True, but neither are they made up of mathematical equations, and it's a lot easier for most people to visualize the deformation of springs and dashpots than the solutions to equations.

A word is needed about the meaning of the term *linear*. For our purposes, a linear response will be defined as one in which the *ratio* of overall stress to overall strain (i.e., the overall modulus) is a function of *time only* and not of the magnitudes of stress or strain,

$$G = \frac{\tau}{\gamma} = G\,(t\ \text{only}) \text{ for linear response} \qquad (18\text{-}1)$$

The Hooken spring responds instantaneously to reach an equilibrium strain γ upon application of a stress, and the strain remains constant as long as the stress is maintained. Sudden removal of the stress results in instantaneous recovery of the strain (Figure 18.2). Doubling the stress on the spring simply doubles the resulting strain, so the spring is linear according to (18-1). (In assuming that the spring instantaneously reaches an equilibrium strain under the action of a suddenly applied constant stress, we have neglected inertial effects. Although it is not necessary to do so, including them would contribute little to the present discussion.) If a stress is suddenly applied to the dashpot, the strain increases with time according to $\gamma = (\tau/\eta)t$ (considering the strain to be zero when the stress is applied) (Figure 18.3). Doubling the stress doubles the slope of the strain-time line, and, at any time, the modulus $G = \tau/\gamma = \eta/t = G\,(t\ \text{only})$. Therefore, the dashpot is also linear.

It may be shown that *any combination of linear elements must be linear*, so any models based on these linear elements, no

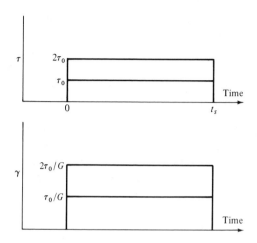

Figure 18.2. Response of spring.

matter how complex, can only represent linear response. Just how realistic is linear response? Well, for most polymers at strains greater than a percent or so (or rates of strain greater than 0.1 sec^{-1}), it's not a good quantitative description. Moreover, even within the *limit of linear viscoelasticity*, it

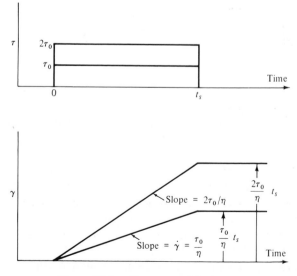

Figure 18.3. Response of dashpot.

usually requires a large number of linear elements (springs and dashpots) to provide an accurate quantitative description of response. Hence, the quantitative applicability of linear models is limited, but they are extremely valuable in visualizing viscoelastic response and in understanding how and why variations in molecular structure influence that response.

The Maxwell Element James Clerk Maxwell realized that neither a linear viscous element (dashpot) nor a linear elastic element (spring) was sufficient to describe his experiments on the deformation of asphalts, so he proposed a simple series combination of the two, the *Maxwell element* (Figure 18.1c). In a Maxwell element, the spring and dashpot support the same stress, so

$$\tau = \tau_{spring} = \tau_{dashpot}. \tag{18-2}$$

Further, the overall strain of the element is the sum of the strains in the spring and dashpot:

$$\gamma = \gamma_{spring} + \gamma_{dashpot}. \tag{18-3}$$

After differentiating (18-3) with respect to time,

$$\dot{\gamma} = \dot{\gamma}_{spring} + \dot{\gamma}_{dashpot}. \tag{18-4}$$

Realizing that $\dot{\gamma}_{dashpot} = \tau/\eta$ and $\dot{\gamma}_{spring} = \dot{\tau}/G$, plugging in and rearranging gives the differential equation for the Maxwell element:

$$\tau = \eta\dot{\gamma} - \left(\frac{\eta}{G}\right)\dot{\tau} = \eta\dot{\gamma} - \lambda\dot{\tau} \tag{18-5}$$

The quantity $\lambda = \eta/G$ has the dimensions of time and is known as a *relaxation time*. Its physical significance will be apparent shortly.

Creep Testing Let's examine the response of the Maxwell element in two mechanical tests commonly applied to polymers. First consider a *creep* test, in which a *constant stress* is instantaneously (or at least very rapidly) *applied* to the material, and the resulting *strain is followed as a function of time*. Deformation after removal of the stress is known as *creep recovery*. As shown in Figure 18.4, the sudden application of

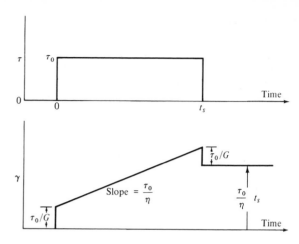

Figure 18.4.   Creep response of Maxwell element.

stress to a Maxwell element causes an instantaneous stretching of the spring to an equilibrium value of τ_o/G, where τ_o is the (constant) applied stress. The dashpot extends linearly with time with a slope of τ_o/η, and it will continue to do so as long as the stress is maintained. Thus, *the Maxwell element is a fluid*, because it will continue to deform as long as it is stressed. When the stress is released, the spring immediately contracts by an amount equal to its original extension, a process known as *elastic recovery*. The dashpot, of course, does not recover, leaving a *permanent set* of $(\tau/\eta)t_s$, the amount the dashpot has extended during the application of stress.

Although real materials never show sharp breaks in a creep test as the Maxwell element does, the model does exhibit the phenomena of elastic strain, creep, recovery, and permanent set, which are often observed with real materials.

Stress Relaxation   Another important test used to study viscoelastic response is *stress relaxation*. A stress relaxation test consists of suddenly *applying a strain* to the sample and *following the stress as a function of time as the strain is held constant*. When the Maxwell element is strained instantaneously, only the spring can respond initially (for an infinite rate of strain, the resisting force in the dashpot is infinite) to a stress of $G\gamma_o$, where γ_o is the constant applied strain. The extended spring then begins to contract, but the contraction is resisted by the dashpot. The more the spring retracts, the smaller its restoring force becomes, and the rate of retraction drops correspondingly. Solution of the differential equation with $\dot{\gamma} = 0$ and the boundary condition $\tau = G\gamma_o$ at $t = 0$ shows that the stress undergoes a first-order exponential decay,

$$\tau = G\gamma_o e^{-t/\lambda}. \tag{18-6}$$

Thus, the relaxation time λ is the time constant for the exponential decay—i.e., the time required for the stress to decay by a factor of $1/e$ or 37 percent. The stress asymptotically drops to zero as the spring approaches complete retraction (Figure 18.5).

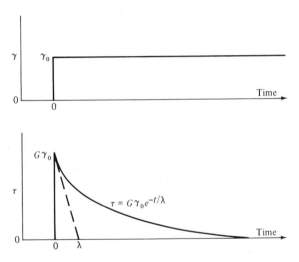

Figure 18.5.   Stress relaxation of Maxwell element.

Stress-relaxation data for linear polymers actually look like the curve for the Maxwell element. Unfortunately, they can't often be fitted quantitatively with a single value of G and a single value of λ; i.e., the decay is not really first order.

Example 1. Examine the response of a Maxwell element in an engineering stress-strain test—i.e., one in which the rate of tensile strain is maintained (approximately) constant at $\dot{\epsilon}_o$.

Solution. Equation (18-5) can be rewritten in tensile notation as

$$\sigma = \eta_e \dot{\epsilon}_o - \left(\frac{\eta_e}{E}\right)\dot{\sigma} \text{ or } \sigma + \left(\frac{\eta_e}{E}\right)\frac{d\sigma}{dt} = \eta_e \dot{\epsilon}_o = \text{constant.}$$

The solution to this differential equation with the boundary condition $\sigma = 0$ at $t = 0$ is

$$\sigma = \eta_e \dot{\epsilon}_o [1 - e^{-(E/\eta_e)t}].$$

Since $d\epsilon/dt = \dot{\epsilon}_o = \text{constant}$, and $\epsilon = 0$ when $t = 0$

$$\epsilon = \dot{\epsilon}_o t.$$

The stress-strain curve is then

$$\sigma = \eta_e \dot{\epsilon}_o [1 - e^{-(E/\eta_e \dot{\epsilon}_o)\epsilon}].$$

This response is sketched in Figure 18.6. Note that, at any given strain, σ increases with the *rate of strain* $\dot{\epsilon}_o$; i.e., the material appears stiffer (has a higher modulus).

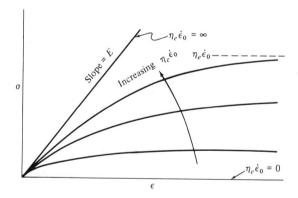

Figure 18.6. Response of a Maxwell element under constant rate of tensile strain (example 1).

As crude as this model might prove to be in fitting actual experimental data, it does account, at least qualitatively, for many of the observed properties of linear polymers in engineering stress-strain tests.

The Voigt-Kelvin Element   If a series combination of a spring and dashpot has its drawbacks, the next logical thing to try is a parallel combination, a Voigt or Voigt-Kelvin element (Figure 18.1d). Here, it is assumed that the crossbars supporting the spring and dashpot always remain parallel—i.e., that *the strain in each element is the same*

$$\gamma = \gamma_{spring} = \gamma_{dashpot} \qquad (18\text{-}7)$$

and that *the stress* supported by the element *is the sum of the stresses in the spring and the dashpot*

$$\tau = \tau_{spring} + \tau_{dashpot}. \qquad (18\text{-}8)$$

Combination of (18-7) and (18-8) with the equations for the deformation of the spring and dashpot gives the differential equation for the Voigt-Kelvin element:

$$\tau = \eta\dot{\gamma} + G\gamma. \qquad (18\text{-}9)$$

When the stress is suddenly applied in a creep test, only the dashpot offers an initial resistance to deformation, so the initial slope of the strain versus time curve is τ_o/η. As the element is extended, the spring provides an increasingly greater resistance to further extension, and so the rate of creep decreases. Eventually, the system comes to equilibrium with the spring alone supporting the stress (with the rate of strain zero, the resistance of the dashpot is zero). The equilibrium strain is simply τ_o/G. Quantitatively, the response is an exponential rise,

$$\gamma = \frac{\tau_o}{G}[1 - e^{-t/\lambda}]. \qquad (18\text{-}10)$$

If the stress is removed after equilibrium has been reached, the strain decays exponentially,

$$\gamma = \frac{\tau_o}{G}e^{-t/\lambda}. \qquad (18\text{-}11)$$

Note that the Voigt-Kelvin model does not continue to deform as long as stress is applied, and it does not exhibit any permanent set (Figure 18.7). It therefore represents a viscoelastic *solid* and gives a fair qualitative picture of the creep response of some crosslinked polymers.

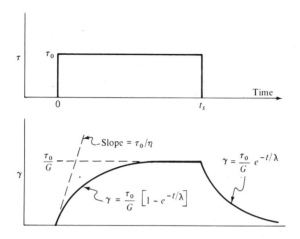

Figure 18.7. Creep response of a Voigt-Kelvin element.

The Voigt-Kelvin model is not suited for representing stress relaxation. The application of an instantaneous strain would be met by an infinite resistance in the dashpot and would therefore require the application of an infinite stress which is obviously unrealistic.

The Three-Parameter Model The next step in the development of linear viscoelastic models is the so-called three-parameter model (Figure 18.1e). The differential equation for this model may be written in operator form as

$$\left[1 + \lambda_1 \frac{d}{dt}\right]\tau = \eta_1 \left[1 + \lambda_2 \frac{d}{dt}\right]\dot{\gamma} \tag{18-12}$$

where $\lambda_1 = (\eta_1 + \eta_2)/G_2$ and $\lambda_2 = \eta_2/G_2$. This model has formed the basis for some quantitative descriptions of material behavior. General quantitative application of this model in three dimensions requires some steps beyond the simple solution of (18-12). First, the stress and rate of strain must be replaced by the stress and rate-of-strain tensors. Second, the model must be applied in convected *material coordinates*, which must then be referred to fixed coordinates to describe the overall deformation; i.e., the model describes the deformation of a differential element of fluid, but this element is in general moving relative to fixed coordinates; thus, to describe the overall deformation of a finite mass of material relative to fixed coordinates, the point deformation and bulk motion must be superimposed. These approaches have met with some limited success as constitutive equations. Further, the form of (18-12) suggests modification by adding higher-order derivatives and more constants

$$[1 + \lambda_1 \frac{d}{dt} + \xi_1 \frac{d^2}{dt^2} + \cdots] \tau = \eta_1 [1 + \lambda_2 \frac{d}{dt} + \xi_2 \frac{d^2}{dt^2} + \cdots] \dot{\gamma}$$

$$(18\text{-}13)$$

This, of course, will fit data to any desired degree of accuracy if enough terms are used.

18.3 THE FOUR-PARAMETER MODEL AND MOLECULAR RESPONSE

The four-parameter model (Figure 18.1f) is a series combination of a Maxwell element with a Voigt-Kelvin element, and its creep response is the sum of their individual responses, as summarized in Figure 18.8. It provides at least a qualitative representation of all the phenomena generally observed with viscoelastic materials: instantaneous elastic strain, retarded elastic strain, equilibrium viscous flow, instantaneous elastic recovery, retarded elastic recovery and permanent set. Of equal importance is the fact that the model parameters can be identified with the various molecular response mechanisms in polymers, and it can therefore be used to predict the influences that changes in molecular structure will have on mechanical properties. The following analogies may be drawn:

a. Dashpot 1 represents molecular slippage. This slip of polymer molecules past one another is responsible for flow. The value of η_1 alone (molecular friction in slip) governs the equilibrium flow of the material.

b. Spring 1 represents the elastic straining of bond angles and lengths. All bonds in polymer chains have equilibrium angles and lengths. The value of G_1 characterizes the resistance to deformation from these equilibrium values. Since these

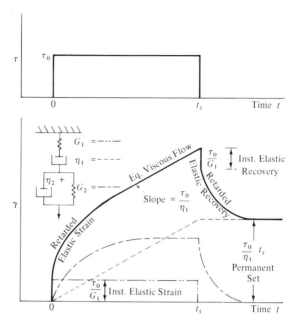

Figure 18.8.   Creep response of a four-parameter model.

deformations involve interatomic bonding, they occur essentially instantaneously from a macroscopic point of view. This type of elasticity is known thermodynamically as *energy elasticity*, because the straining of bond angles and lengths increases the material's internal energy (see chapter 14).

c. Dashpot 2 represents the resistance of the polymer chains to uncoiling and coiling, caused by temporary mechanical entanglements of the chains and molecular friction during these processes. Since coiling and uncoiling require cooperative motion of many chain segments, they cannot occur instantaneously and hence account for retarded elasticity.

d. Spring 2 represents the restoring force brought about by the thermal agitation of the chain segments, which tends to return chains oriented by a stress to their most random or highest entropy configuration. This type of restoring force is known, therefore, as *entropy elasticity* (chapter 14).

Example 2. By using the four-parameter model as a basis, sketch qualitatively the effects of (a) increasing molecular weight and (b) increasing degrees of crosslinking on the creep response of a linear, amorphous polymer.

Solution. a. As discussed in chapter 15, the equilibrium zero-shear (linear) viscosity of polymers, represented by η_1 in the model, increases with the 3.4 power of $\bar{M}_w$. Thus, the slope in the equilibrium flow region, τ_o/η, is greatly decreased as the molecular weight increases, and the permanent set, $(\tau_o/\eta_1)t_s$, is reduced correspondingly (Figure 18.9).

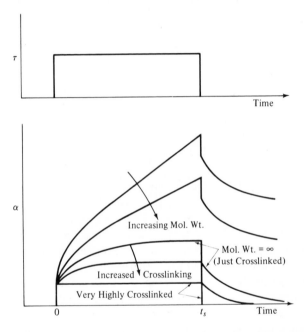

Figure 18.9. The effects of molecular weight and crosslinking on the creep response of an amorphous polymer.

b. Light crosslinking represents the limit of case (a) above, when the molecular weight reaches infinity, as all the chains are hooked together by crosslinks. Under these conditions, they *can't* slip past one another, so η_1 becomes infinite. If the crosslinking is light (the crosslinks few and far between), as in a rubber band, coiling and uncoiling won't be hindered appreciably. Note that crosslinking converts the material from a fluid to a solid (it eventually reaches an equilibrium strain under the application of a constant stress), and it eliminates permanent set. The equilibrium modulus will be on the order of 10^6 to 10^7 dynes/cm^2, the characteristic rubbery modulus.

Further crosslinking begins to hinder the ability of the chains to uncoil and raises the restoring force (increases η_2 and G_2). At high degrees of crosslinking, as in hard rubber (ebonite), the only response mechanism left is straining bond angles and lengths, giving rise to an almost perfectly elastic material with a modulus on the order of 10^{10} to 10^{11} dynes/cm^2, the characteristic glassy modulus.

The magnitude of t_s on the diagram will of course depend on the applied stress and on the particular material and its temperature—i.e., on the values of the model parameters. Well below T_g, for example, where η_1 and η_2 are very large, t_s might be on the order of days or weeks in order that appreciable retarded elasticity and flow be observed. Above T_g, it might represent seconds or less. The important thing to keep in mind, however, is that designs based on short-term property measurements will be inadequate if the object must support a stress for longer periods of time.

The four-parameter model nicely accounts for the interesting examples of viscoelastic response mentioned earlier. For example, dashpot 1 allows viscous flow, while the elastic restoring forces of springs 1 and 2 provide the rubber band elasticity responsible for the Weissenberg effect. In engineering stress-strain tests, the moduli of polymers are observed to increase with the applied rate of strain. At high rates of strain, spring 1 provides the major response mechanism. As the rate of strain is lowered, dashpot 1 and the Voigt-Kelvin element contribute more and more to the overall deformation, giving a greater strain at any stress—i.e., a lower modulus. When silly putty is bounced (stress applied rapidly for a short period of time), spring 1 again provides the major response mechanism. There isn't time for appreciable flow of the dashpots 1 and 2; i.e., not much of the initial potential energy is converted to heat through the molecular friction involved in slippage and uncoiling, and the material behaves in an almost perfectly elastic fashion. When it's stuck on the wall, the stress—in this case due to its own weight—is applied for a long period of time, and it flows downward as a result of the molecular slip represented by dashpot 1.

18.4   VISCOUS OR ELASTIC RESPONSE?
THE DEBORAH NUMBER (7)

Thus, whether a viscoelastic fluid behaves as an elastic solid or a viscous liquid depends on the relation between the time scale of the experiment and the time required for the time-dependent mechanisms to respond. Strictly speaking, the concept of a single relaxation time applies only to first-order response and thus is not applicable to real materials, in general. Nevertheless, *a characteristic time, λ_c, for any material can always be defined* as, for example, the time required for the material to reach $1/e$ of its ultimate elastic response to a step change. The ratio of this characteristic time to the time scale of the experiment is the *Deborah number*:

$$N_{De} = \frac{\lambda_c}{t_s}. \qquad (18\text{-}14)$$

At high Deborah numbers, response will be elastic; at low Deborah numbers, it will be viscous. ("The mountains flowed before the Lord"—from the Song of Deborah, Book of Judges V. For the Lord, t_s is exceedingly large (7).)

At Deborah numbers approaching zero (i.e., in equilibrium viscous flow), the elastic response mechanisms have reached their equilibrium strain under the applied stress, and deformation is due solely to molecular slippage, dashpot 1. Under these conditions, materials can be treated by the techniques outlined in chapter 16 for purely viscous fluids. Modifications of the devices described can be used to obtain information on the material's elastic properties. For example, if the stress is suddenly removed from a rotational viscometer, the creep recovery or elastic recoil of the material can be followed. This provides a value of λ_c for the material.

18.5   QUANTITATIVE
APPROACHES (1, 2, 3, 4)

Although the four-parameter model is so useful from a conceptual standpoint, the four parameters usually aren't sufficient to provide an accurate fit of most experimental data.

To do so, and to infer some detailed information about molecular response, more general models have been developed. The *generalized Maxwell model* (Figure 18.10) is used to describe

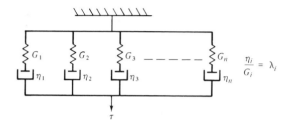

Figure 18.10.   Generalized Maxwell model.

stress-relaxation experiments.   The stress relaxation of an individual Maxwell element is given by

$$\tau_i(t) = \gamma_o G_i e^{-t/\lambda_i} \qquad (18\text{-}15)$$

where $\lambda_i = \eta_i/G_i$. The relaxation of the generalized model, in which the individual elements are all subjected to the same constant strain, γ_o, is then

$$\tau(t) = \gamma_o \sum_{i=1}^{n} G_i e^{-t/\lambda_i}. \qquad (18\text{-}16)$$

Expressed in terms of a time-dependent modulus, $G(t)$, the response is

$$G(t) = \frac{\tau(t)}{\gamma_o} = \sum_{i=1}^{n} G_i e^{-t/\lambda_i}. \qquad (18\text{-}17)$$

Now, if n is large, the summation of individual, discrete moduli in (18-17) may be approximated by the integral of a *continuous distribution* of *relaxation times*, $G(\lambda)$:

$$G(t) = \int_o^\infty G(\lambda) e^{-t/\lambda} \, d\lambda. \qquad (18\text{-}18)$$

Note that, while the G_i have units of dynes/cm^2, $G(\lambda)$ is in dynes/cm^2 sec.   Note also that, if the generalized Maxwell model is to represent a viscoelastic solid (e.g., a crosslinked polymer), at least one of the viscosities has to be infinite.

For creep tests, a generalized Voigt-Kelvin model is used (Figure 18.11). The creep response of an individual Voigt-

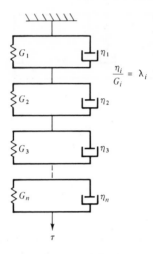

Figure 18.11. Generalized Voigt-Kelvin model.

Kelvin element is given by

$$\gamma_i(t) = \tau_o J_i(1 - e^{-t/\lambda_i}) \qquad (18\text{-}19)$$

where $J_i = 1/G_i$ is a *compliance* (cm^2/dyne). The response of the array, in which each element is subjected to the same constant applied stress, τ_o, is then

$$\gamma(t) = \tau_o \sum_{i=1}^{n} J_i(1 - e^{-t/\lambda_i}) \qquad (18\text{-}20)$$

or, in terms of a time-dependent compliance, $J(t)$,

$$J(t) = \frac{\gamma(t)}{\tau_o} = \sum_{i=1}^{n} J_i(1 - e^{-t/\lambda_i}). \qquad (18\text{-}21)$$

Again, for large n, the discrete summation above may be approximated by

$$J(t) = \int_0^\infty J(\lambda)(1 - e^{-t/\lambda})\, d\lambda \qquad (18\text{-}22)$$

where $J(\lambda)$ is the *continuous distribution of retardation times* ($cm^2/dyne\ sec$).

If the generalized Voigt-Kelvin model is to represent a visco-elastic liquid (e.g., a linear polymer), the modulus of one of the springs must be zero (infinite compliance), leaving a simple dashpot in series with all the other Voigt-Kelvin elements. Sometimes, the steady-flow response of this lone dashpot, $\gamma_{dashpot} = (\tau_o/\eta)t$, is subtracted from the overall response, leaving the compliances to represent only the elastic contributions to the overall response; i.e.,

$$\gamma(t) = \frac{\tau_o}{\eta} t + \tau_o \sum_{i=1}^n J_i^\dagger (1 - e^{-t/\lambda_i}) \qquad (18\text{-}23)$$

$$J(t) = \frac{\gamma(t)}{\tau_o} = \frac{t}{\eta} + \sum_{i=1}^n J_i^\dagger (1 - e^{-t/\lambda_i}) \qquad (18\text{-}24)$$

$$J^\dagger(t) = J(t) - \frac{t}{\eta} = \sum_{i=1}^n J_i^\dagger (1 - e^{-t/\lambda_i}) \qquad (18\text{-}25)$$

$$J^\dagger(t) = J(t) - \frac{t}{\eta} = \int_0^\infty J^\dagger(\lambda)(1 - e^{-t/\lambda})\, d\lambda. \qquad (18\text{-}26)$$

Here, the daggers indicate that the purely viscous flow has been removed and is treated separately.

Procedures are available for extracting $G(\lambda)$ or $J(\lambda)$ from experimental data. Once known, they can be used in (18-18) or (18-22) for predicting creep or stress relaxation response in the linear region. So what, if you first had to measure the response to determine these functions? Well, for one thing, $G(\lambda)$ can in principle be obtained from $J(\lambda)$ and vice-versa, if one distribution is known over the range $0 < \lambda < \infty$. Although the functions are never obtainable over the complete range, reasonable approximations are available. Thus, creep response can be predicted from stress-relaxation measurements, and vice-versa. This interconvertability applies to a variety of mechanical responses in addition to the two types discussed here. Furthermore, the shape of the distributions provides the polymer scientist with information on molecular response mechanisms within the polymer. For example, peaks in a certain region of λ might imply motion of side chains on the molecules. This type of information can lead to the design of polymers with the type of side chains needed to provide particular mechanical properties.

18.6 DYNAMIC MECHANICAL TESTING (1, 4)

Creep and stress-relaxation measurements correspond to the use of step-response techniques to analyze the dynamics of electrical and process systems. Those familiar with these areas know that frequency-response analysis is perhaps a more versatile tool for investigating system dynamics. An analogous procedure, *dynamic mechanical testing*, is applied to the mechanical behavior of viscoelastic materials. It is based on

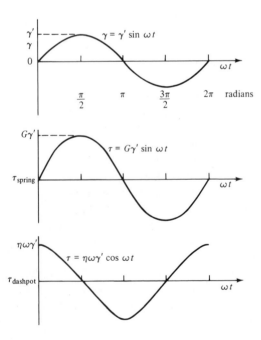

Figure 18.12.   Stress in a linear spring and a linear dashpot in response to a sinusoidal applied strain.

the fundamentally different response of viscous and elastic elements to a sinusiodally varying stress or strain.

If a sinusiodal strain, $\gamma = \gamma' \sin \omega t$, (where ω is the angular frequency, radian/sec) is applied to a linear spring, since $\tau = G\gamma$, the resulting stress, $\tau = G\gamma' \sin \omega t$, is *in phase with the strain.* For a linear dashpot, however, because the stress is proportional to the *rate of strain* rather than the strain, $\tau = \eta\dot{\gamma} = \eta\omega\gamma' \cos \omega t$, the *stress is 90° out of phase with the strain.* These relations are sketched in Figure 18.12.

As might be expected, viscoelastic materials exhibit some sort of intermediate response, which might look like Figure 18.13b. This can be thought of as being a projection of two vectors, τ^* and γ^*, rotating in the complex plane (Figure 18.13a). The angle between these vectors is the *phase angle,* δ ($\delta = 0$ for a purely elastic material and 90° for a purely viscous material). It is customary to resolve the vector repre-

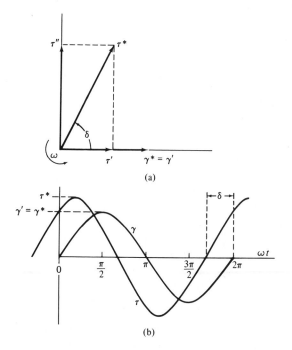

Figure 18.13.   Quantities in dynamic testing. (a) rotating vector diagram, (b) stress and strain.

senting the dependent variable into components in phase (designated by ') and 90° out-of-phase (designated by '') with the independent variable. In this example, the applied *strain* is the independent variable, so the observed stress vector (τ^*) is resolved into its in-phase (τ') and out-of-phase (τ'') components, $\gamma^* = \gamma'$, and $\gamma'' = 0$.

An *in-phase* or *storage modulus* is defined by

$$G' = \frac{\tau'}{\gamma'} \qquad \text{storage modulus (in-phase component)} \qquad (18\text{-}27)$$

and an *out-of-phase* or *loss modulus* is defined by

$$G'' = \frac{\tau''}{\gamma'} \qquad \text{loss modulus (out-of-phase component).} \qquad (18\text{-}28)$$

A *complex modulus*, G^*, may then be defined as the *vector sum* of the in-phase and out-of-phase moduli (using i to denote an out-of-phase unit vector):

$$G^* = G' + iG'' = \frac{\tau' + i\tau''}{\gamma'} = \frac{\tau^*}{\gamma^*} \qquad \text{complex modulus.} \qquad (18\text{-}29)$$

From the geometry of the situation, it may be seen that the *loss tangent*, tan δ, is

$$\tan \delta = \frac{\tau''}{\tau'} = \frac{G''}{G'} \qquad \text{loss tangent.} \qquad (18\text{-}30)$$

Since G', G'' and tan δ are related by (18-30), only two of the three are independent and need be reported.

Example 3. Obtain the quantities τ', τ'', G', G'', tan δ and G^* for a Voigt-Kelvin element subjected to the strain $\gamma = \gamma' \sin \omega t$.

Solution. The differential equation of the Voigt-Kelvin element is $\tau = G\gamma + \eta\dot\gamma$ (18-9).

$$\dot\gamma = \omega\gamma' \cos \omega t$$

$$\tau = G\gamma' \sin \omega t + \eta\omega\gamma' \cos \omega t$$

The sin term is the stress component *in phase* with the strain and the cos term is 90° out-of-phase. Therefore,

$$\tau' = G\gamma' \quad \text{and} \quad \tau'' = \eta\omega\gamma'$$

$$G' = \frac{\tau'}{\gamma'} = \frac{G\gamma'}{\gamma'} = G \quad \text{and} \quad G'' = \frac{\tau''}{\gamma'} = \eta\omega$$

$$\tan \delta = \frac{\tau''}{\tau'} = \frac{G''}{G'} = \frac{\eta\omega}{G}$$

$$G^* = \sqrt{(G')^2 + (G'')^2} = \sqrt{G^2 + \eta^2\omega^2}.$$

The latter result shows how the apparent stiffness of the element, G^*, increases with frequency. This can be deduced qualitatively from the model. At low frequencies ($\omega \to o$), there is no resistance in the dashpot, and $G^* \to G$. At high frequencies ($\eta\omega \gg G$), the resistance of the dashpot predominates, and $G^* \to \eta\omega$.

What is the physical significance of the quantities just defined? This can best be appreciated by considering what happens to the energy applied to a sample undergoing cyclic deformation. The work done on a unit volume of material undergoing a pure shear deformation is

$$w = \int \tau \, d\gamma \left(\frac{\text{ergs}}{\text{cm}^3}\right). \tag{18-31}$$

From Figure 18.13,

$$\gamma = \gamma' \sin \omega t \tag{18-32}$$

and

$$\tau = \tau^* \sin (\omega t + \delta). \tag{18-33}$$

Differentiating (18-32) with respect to (ωt) gives

$$d\gamma = \gamma' \cos \omega t \, d(\omega t). \tag{18-34}$$

After inserting (18-33) and (18-34) into (18-31),

$$w = \tau^* \gamma' \int \sin (\omega t + \delta) \cos \omega t \, d(\omega t). \tag{18-35}$$

Let's first consider the work done on the first quarter-cycle of applied strain; i.e., integrate (18-35) between 0 and $\pi/2$. Using appropriate trigonometric identities and a good set of integral tables gives

$$w \left(1\text{st} \frac{1}{4} \text{cycle}\right) = \tau^* \gamma' \left(\frac{\cos \delta}{2} + \frac{\pi}{4} \sin \delta\right). \tag{18-36}$$

Put in terms of moduli by using the trigonometry of Figure 18.13,

$$w \left(1\text{st} \frac{1}{4} \text{cycle}\right) = \frac{(\gamma')^2}{2} G' + \frac{\pi}{4} (\gamma')^2 G''. \qquad (18\text{-}37)$$

The first term on the right-hand side of (18-37) is simply the work done in straining a Hookean spring of modulus G' by an amount γ'. It therefore represents the energy stored elastically in the material during its straining in the first quarter cycle. Hence, G' is the storage modulus. If the applied mechanical energy (work) is not stored elastically, it must be lost—converted to heat through molecular friction (i.e., viscous dissipation within the material). This is precisely what the second term on the right represents, so G'' is known as the loss modulus.

When we consider the second quarter of the cycle, we find that integrating (18-35) from $\pi/2$ to π gives results identical to (18-37) *except that the sign on first (storage) term is negative.* This simply means that the energy stored elastically in straining the material from 0 to γ' is recovered when it returns from γ' to 0. Thus, over a half-cycle or a full cycle, there is no net work done or energy lost through the elastic component. The sign of the second term, however, is positive for any quarter cycle, so the net energy loss (converted to heat within the material) for a full cycle (also obtainable by integrating (18-35) between 0 and 2π) is simply

$$w \text{(complete cycle)} = \pi (\gamma')^2 G'' = \pi (\gamma')^2 G' \tan \delta. \qquad (18\text{-}38)$$

These results are of direct importance in the design of polymeric objects which are subjected to cyclic deformation. In a tire, for example, high temperatures contribute to rapid degradation and wear. It is therefore desirable to choose a rubber compound with as low a G'' as possible to minimize energy dissipation and the resultant heat buildup. In the design of an engine mount, however, the object is usually to prevent the transmission of vibration from the engine. Here, a material with a large G'' would dissipate the vibrational energy as heat rather than transmitting it.

Basically, two methods are available for determining dynamic properties. Forced-oscillation devices apply a known sinusoidal stress or strain and measure the resulting strain or stress. Generally, they are designed to study the variation of G' and G'' (or tan δ) over ranges of both temperature and angular frequency ω. The functions $G'(\omega)$ and $G''(\omega)$ are at least approximately convertible to $G(\lambda)$ and $J(\lambda)$ as obtained from creep and stress-relaxation measurements, and vice-versa.

Free-oscillation measurements are made with a torsion pendulum, in which the sample is given an initial torsional displacement, and the frequency and amplitude decay of the oscillations are observed upon release. Although the frequency can be varied somewhat by changing the moment of inertia of the oscillating portion of the mechanism, torsion pendula are usually used to study only the temperature dependence of G' and G'' at a constant, relatively low frequency.

Figure 18.14 illustrates G' and damping $\cong \pi$ tan δ versus T for polymethyl methacrylate, an amorphous, linear polymer. The data were obtained with a torsion pendulum at about 1 cycle/sec. At low temperatures, the typical glassy modulus 10^{10} to 10^{11} dynes/cm^2 is observed. In the vicinity of 110C to 130C, G' drops precipitously, ultimately reaching a plateau of 10^6 to 10^7 dynes/cm^2, the typical rubbery modulus. Also, a sharp peak in the damping is observed in this region. Although the temperature at which this peak and drop are observed is frequency dependent, at low frequencies (such as are obtained with a torsion pendulum), they are identified with the material's glass transition temperature, and the drop in G' is indicative of the decrease in stiffness at T_g, going from the straining of bond angles and lengths to coiling and uncoiling as the dominant response mechanism. The damping peak represents the onset of cooperative motion of 40–50 main-chain carbon atoms at T_g (chapter 8). In terms of the four-parameter model, below T_g, only spring 1 is operative, and the material is almost completely elastic (low damping). In the vicinity of T_g, the viscosity of dashpot 2 drops to the point where it can deform and dissipate energy, giving the damping peak. At higher temperatures, its viscosity drops to the point where it dissipates little energy, and the material is again highly elastic, mainly through spring 2. At still higher temperatures, the modulus drops off rapidly due to viscous flow, dashpot 1. The broad damping hill centered at about 40C (and the accompanying

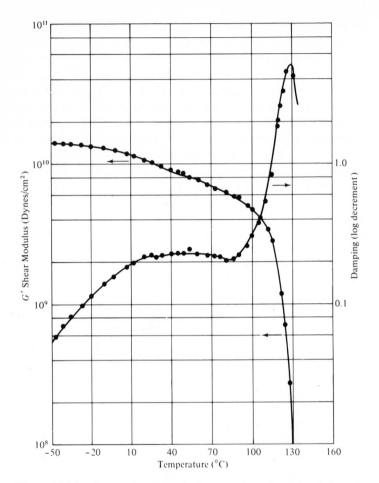

Figure 18.14.   Dynamic mechanical properties of polymethyl methacrylate (6). The data were obtained with a torsion
pendulum at about 1 cycle/sec.

gradual drop in G') in Figure 18.14 has been shown to arise

$$\text{O}$$
$$\|$$

from the motion of the $-\text{C}-\text{O}-\text{CH}_3$ side groups.

Note that the dimensions of the angular frequency, ω, are
$\sec^{-1}$. It thus corresponds to a *reciprocal* time scale. For dynamic (oscillating) deformations, then, the Deborah number is

$$N_{De} = \lambda_c \omega. \tag{18-39}$$

In dynamic tests, a viscoelastic material will become more solid-like as the frequency is increased and the time-dependent response mechanisms become unable to follow the rapidly reversing stress. In the limit of very high frequencies, the straining of bond angles and lengths will be the only operative response mechanism, and the polymer will exhibit the typical glassy modulus even though it may be well above its glass-transition temperature.

18.7 TIME-TEMPERATURE SUPERPOSITION

Anyone who has ever wrestled with a cheap garden hose in cold weather appreciates the fact that polymers become stiffer at lower temperatures; i.e., at high temperatures they are softer or more liquid-like, and when cooled they become stiffer or more solid-like. In preceding examples, we have seen that the time scale (or frequency) of the application of stress has a similar influence on mechanical properties, short times (or high frequencies) corresponding to low temperatures and long times (low frequencies) to high temperatures. A ball of silly putty, if heated sufficiently above room temperature (assuming it didn't degrade first), would simply stay on the floor like clay when dropped. Conversely, if cooled sufficiently below room temperature and stuck on the wall, it would exhibit mainly elastic deformation for long periods of time.

The quantitative application of this idea, *time-temperature superposition*, is one of the most important principles of polymer physics. It is based on the fact that the Deborah number determines quantitatively just how a viscoelastic material will behave mechanically. Changing either t_s (or ω) *or* λ_c can change N_{De}. The nature of the applied deformation determines t_s (or ω), *while a polymer's characteristic time is a function of temperature.* The higher the temperature, the more thermal energy the chain segments possess, and the more rapidly they are able to respond, thus lowering λ_c.

Although time-temperature superposition is applicable to the widest variety of viscoelastic response tests (creep, dynamic, etc.), it is usually illustrated with stress relaxation, since much of the early development was done with stress relaxation. Figure 18.15 illustrates tensile stress relaxation data at various

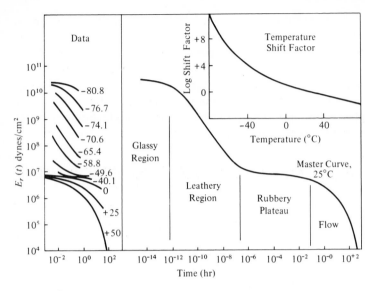

Figure 18.15.   Time-temperature superposition for NBS polyisobutylene (adapted from 5).

temperatures for polyisobutylene, plotted in the form of a time-
dependent tensile (Young's) modulus, $E_r(t)$ versus time on a
log-log scale.

$$E_r(t) = \frac{f(t)/A}{\Delta l/l \,(\text{constant})} \qquad (18\text{-}40)$$

where $f(t)$ is the measured tensile force in the sample, A its
cross-sectional area, and l its length.   The technique is, of
course, equally applicable to shear deformation.   In stress relaxation, the lower measurement time limit is set by the assumption that the constant strain is applied instantaneously.
In practice, inertia and other mechanical limitations make this
impossible, so data are only valid at times a couple of orders of
magnitude longer than it actually takes to apply the constant
strain.   The upper limit is set by the dedication of the experimenter and the long-term stability of the sample and equipment.   These data were obtained over a range from seconds to
a couple of days.   As might be expected, the modulus drops
with time at a given temperature, and, at a given time, it drops
with increasing temperature.

Staring at the curves for a while indicates that they appear to be sections of one continuous curve, chopped up, with the sections displaced along the log time axis. That this is indeed so is shown in Figure 18.15. Here, 25C has *arbitrarily* been chosen as a reference temperature, T_o, and the curves for other temperatures have been shifted along the log time axis to line up with it. The data below 25C must be shifted to the left (shorter times) and that above to the right (longer times), giving a *master curve* at 25C.

Sometimes, the relaxation moduli at each temperature T are corrected to the reference temperature T_o by multiplying by the ratio T_o/T before superposing. This correction is based on the theory of ideal rubber elasticity (chapter 14) which states that the modulus is proportional to the absolute temperature. This procedure is open to question, however. While ideal rubber elasticity might be a reasonable approximation when the major response mechanism is chain coiling and uncoiling, it certainly isn't where response is dominated by the straining of bond angles and lengths (glassy region) or molecular slippage (viscous flow region).

Shifting a curve along the log time axis corresponds to multiplying every value of its absicca by a constant factor (it is immaterial what kind of scale is used for the ordinate). This constant factor, which brings a curve at a particular temperature into alignment with the one at the reference temperature, is known as the *temperature shift factor, a_T,*

$$a_T = t_T/t_{T_o} \quad \text{(for same response)} \quad (18\text{-}41)$$

where t_T is the time required to reach a particular response (modulus, in this case) at temperature T and t_{T_o} is the time required to reach the *same response* at the reference temperature, T_o. For temperatures above the reference temperature, it takes less time to reach a particular response (the material responds faster; i.e., it has a shorter relaxation time), so a_T is less than one, and vice-versa. The logarithm of the experimentally determined temperature shift factor is plotted as a function of temperature in Figure 18.15.

The master curve now represents stress relaxation at 25C *over fourteen decades of time.* Multiplication by the appropriate value of a_T establishes the master curve at any other temperature and can thus be used to predict response at that temperature over the entire time scale.

There are two additional aspects which add to the utility of the time-temperature superposition concept. First, the same temperature shift factors apply to a particular polymer regardless of the nature of the mechanical response; i.e., the shift factors as determined in stress relaxation are applicable to the prediction of the time-temperature behavior in creep or dynamic testing. Second, *if the polymer's glass-transition temperature is chosen as the reference temperature*, the shift factors are given, to a good approximation, by the Williams-Landel-Ferry (WLF) equation in the range $T_g < T < (T_g + 100C)$.

$$\log a_T = \frac{-17.44(T - T_g)}{51.6 + (T - T_g)} \quad \text{(for } T_o = T_g\text{)} \quad (18\text{-}41)$$

Example 4. The damping peak for polymethyl methacrylate in Figure 18.14 is located at 130C. Assuming the data were obtained at a frequency of 1 cycle/sec, where would the peak be located if measurements were made at 1000 cycles/sec? For polymethyl methacrylate, $T_g = 105C$.

Solution. Keep in mind that frequency is a *reciprocal* time scale, and apply the WLF equation; then

$$\log a_T = \log \frac{\omega_{T_g}}{\omega_T} = \frac{-17.44(T - T_g)}{51.6 + T - T_g}.$$

Shift the measurements at 1 cps to T_g (i.e., find the frequency at which the peak would be located at T_g); then

$$\log \frac{\omega_{105°}}{\omega_{130°}} = \frac{-17.44(130 - 105)}{51.6 + 130 - 105} = -5.69$$

$$\frac{\omega_{105°}}{\omega_{130°}} = 2.04 \times 10^{-6}$$

and

$$\omega_{105°} = (1 \text{ cps})(2.04 \times 10^{-6}) = 2.04 \times 10^{-6} \text{ cps}$$

now, shift from T_g to T,

$$\log \frac{2.04 \times 10^{-6}}{10^3} = -8.69 = \frac{-17.44(T - 105)}{51.6 + T - 105}$$

$$T = 156°C$$

This illustrates quantitatively the statements made at the end of section 18.6.

Example 5. The master curve for the polyisobutylene in Figure 18.15 indicates that stress relaxes to a modulus of 10^6 dynes/cm^2 in about 10 hours *at 25C*. By using the WLF equation, estimate the time it will take to reach the same modulus at a temperature of $-20C$. For PIB, $T_g = -70C$.

Solution. To use the WLF equation, the reference temperature must be $T_g = -70C$.

$$\log\left(\frac{t_T}{t_{T_g}}\right) = \log\frac{t_{25°}}{t_{-70°}} = \frac{-17.44\,(25\,+\,70)}{51.6\,+\,25\,+\,70} = -11.3$$

$$\frac{t_{25°}}{t_{-70°}} = 5.01 \times 10^{-12} \qquad \text{and}$$

$$t_{-70°} = \frac{10}{5.01 \times 10^{-12}} = 2 \times 10^{12}\,\text{hours}$$

$$\log\frac{t_{-20°}}{t_{-70°}} = \frac{-17.44\,(-20\,+\,70)}{51.6\,-\,20\,+\,70} = 8.59$$

$$\frac{t_{-20°}}{t_{-70°}} = 2.57 \times 10^{-9}$$

$$t_{-20°} = (2 \times 10^{12})(2 \times 57 \times 10^{-9}) = 5{,}140\,\text{hours}$$

This shows how lowering the temperature maintains mechanical stiffness for much longer periods of time.

Let's take a closer look at the stress-relaxation master curve. At low temperatures or short times, only bond angles and lengths can respond, and so the typical glassy modulus, 10^{10} to 10^{11} dynes/cm^2 is observed. This is the so-called *glassy region*. At longer time or higher termperatures, response is governed by the uncoiling of the chains, with the characteristic modulus 10^6 to 10^7 dynes/cm^2 in the *rubbery plateau*. The intermediate region, where the modulus drops from glassy to rubbery is sometimes known as the *leathery region* from the leather-like feel of polymers in this region. At still longer times, the modulus drops off rapidly as a result of molecular slippage, the *viscous-flow region*. Figure 18.16 illustrates the effect of molecular weight and crosslinking on stress relaxation. Molecular weight should have no influence on straining of bond angles and lengths or on uncoiling, so the glassy and

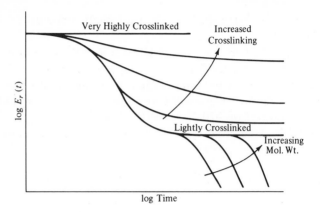

Figure 18.16.   The effects of molecular weight and crosslinking on stress-relaxation master curves.

rubbery moduli are unchanged.   Viscous flow, however, is severely retarded by increasing molecular weight and in the limit of infinite molecular weight (light crosslinking) is eliminated entirely, and the curve levels off with the rubbery modulus.   Higher degrees of crosslinking restrict uncoiling, ultimately leading to a material which responds only by straining of bond angles and lengths.   The effects of crystallinity are similar to those of crosslinking.

To illustrate the effect of temperature on mechanical properties, it is sometimes preferable to plot the property versus temperature for constant values of time.   For example, the data in Figure 18.15 are sometimes crossplotted as $E_r(T)$ versus T at $t = 10$ sec (the ten-second modulus), etc.

REFERENCES

1. Ferry, J. D. *Viscoelastic Properties of Polymers.*   John Wiley & Sons, Inc., New York, 1961.

2. Eirich, F. R., ed. *Rheology*, Vol. 1–4. Academic Press, New York, 1956–1964.

3. Tobolsky, A. V. *Properties and Structure of Polymers.* John Wiley & Sons Inc., New York, 1960.

4. Nielsen, L. E. *Mechanical Properties of Polymers*, chapter 7. Reinhold Publishing Corp., New York, 1962.

5. Tobolsky, A. V. and E. Catsiff, Elastoviscous properties of polyisobutylene (and other amorphous polymers) from stress relaxation studies. IX. A summary of results. *J. Polymer Sci.*, **19,** No. 111, 1956.

6. Nielsen, L. E. Dynamic mechanical properties of polymers. *Soc. Plas. Eng. J.*, **16,** no. 525, 1960.

7. Reiner, M. The Deborah number. *Phys. Today*, Jan., p. 62, 1964.

Section IV

POLYMER TECHNOLOGY

19
Processing

19.1 INTRODUCTION

The term *processing* is used here to describe the technology of converting raw polymer, or compounds containing raw polymer, to articles of a desired shape. Considering the wide variety of polymer types and the even wider variety of articles made from them, a complete description of the myriad of processing techniques which have sprung up through the years would be impossible here. Instead, the more common procedures will be outlined, and their bases in the fundamental material properties previously considered will be discussed.

19.2 MOLDING

Molding consists of forcing a material in the fluid state into a mold under pressure, where it solidifies and takes the shape of the mold cavity.

Injection molding is the most common means of fabricating thermoplastic articles. The molding compound, usually in the form of pellets approximately 1/8-inch cubes or 1/8" diameter × 1/8" long cylinders (molding powder) is fed from a hopper to a heated barrel, where it is heated above the glass transition temperature and/or crystalline melting point. The viscous melt is then forced under pressure into a cold mold where it solidifies, after which the mold opens and the part is ejected.

Figure 19.1 illustrates the older type of injection molding machine. Pellets fall from the hopper into the barrel, which most commonly is electrically heated. A hydraulic ram or plunger pushes the pellets forward. As they move past the

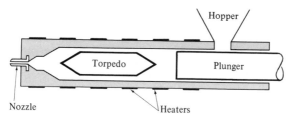

Figure 19.1.   Plunger-type injection molding machine.

heated walls, they are melted and forced to flow around a *torpedo* or *spreader* which promotes heat transfer and homogenization. The molten polymer is then injected through a nozzle into the water-cooled mold, where it travels in turn through a *sprue, runners* and a narrow *gate* into the cavity itself (Figure 19.2). As soon as the gate has frozen off, trapping the material in the cavity, the plunger is retracted to begin another cycle. When the part has cooled sufficiently, the mold opens, and *knockout pins* eject the parts with sprue and runners attached.

The molds are held closed either by hydraulic cylinders or toggle mechanisms. These must be pretty hefty, because injection pressures in the mold can reach the order of 10,000 psi. The molds themselves often involve much intricate hand labor and can thus be quite expensive, but, when amortized over a production run of millions of parts, the contribution of mold cost to the cost of the finished item may be insignificant.

A newer type of injection press, with an in-line screw preplasticizer, is shown in Figure 19.3. Here, the pellets are conveyed forward by a rotating screw which deposits molten plastic at the front end of the barrel. As the material is conveyed forward by the rotating screw, the screw is forced backward while material from the previous shot is cooling in the mold. The cooled parts are then ejected from the mold, the mold closes, and the screw is pushed forward hydraulically, injecting a new shot into the mold. The preplasticizer provides better heat transfer, allowing more material to be plasticized (melted) in a given time, and it gives a more homogeneous melt. Thus machine capacity is greater, and the molded parts have less frozen-in stress.

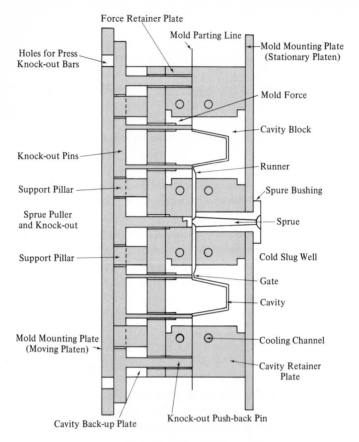

Figure 19.2.   Two-cavity injection mold (Modern Plastics Encyclopedia, McGraw-Hill Book Co., 1969–70 ed.).

Injection-molding presses are rated in terms of tons of mold-
clamping capacity and in ounces (of general-purpose polystyrene) of shot size. In terms of the latter, they range from
1-ounce laboratory units to 500-ounce monsters which are
used to mold such things as garbage cans in a single shot.
They are usually completely automated. Cycle times depend
on plasticizing capacity (the rate at which polymer can be
melted) and the thickness of the part being molded (which
governs cooling time in the mold). Typical values are between
one-half and two minutes. Since a mold may contain many

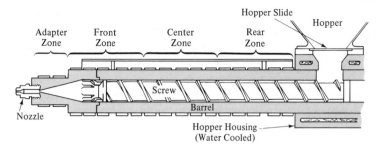

Figure 19.3.   Injection molding machine with in-line screw preplasticizer (Modern Plastics Encyclopedia, McGraw-
Hill Book Co., 1969–70 ed.).

individual cavities, tremendous production rates can be achieved. Sprues and runners are usually removed by hand and may be chopped and recycled back to the hopper.

Until recently, injection molding was confined exclusively to thermoplastics. It is now being used increasingly for thermosets, which are injected into a heated mold where they solidify through the curing (crosslinking) reaction. This is a tricky operation. The compound must be heated just enough in the barrel to achieve fluidity and injected into the mold before it begins to cure appreciably. This requires precise control of temperatures and cycle timing to prevent premature cure in the barrel and the resulting shutdown and cleanout operation. Thermoset sprues and runners cannot be recycled, of course, and must be minimized to cut waste.

Thermosetting compounds are traditionally *compression* molded. The molds are mounted in hydraulic presses on steam, electric or oil heated *platens*. The molding compound is fed to the heated mold, which closes and maintains the material under pressure until cured. The part is then ejected from the mold.

Molding compound in the form of granules or powder may be fed to the mold automatically in weighed shots or as preformed (by cold pressing) tablets. The charge is often preheated to reduce heating time in the mold and, thereby, cycle time.

Mold temperature is the critical variable in compression molding. The higher it is, the faster the material cures, but, if it cures too fast, it will not have enough time to fill thin sections and far corners of the mold so a compromise must be reached. Material suppliers attempt to optimize the cure characteristics to provide minimum cycle times.

Although used occasionally in the laboratory, compression molding is impractical for thermoplastics since the part has to be cooled before ejection from the mold. This requires alternate heating and cooling of a large mass of metal (the mold), which wastes much time and energy.

A variation of compression molding is known as *transfer molding*. Here, the material is melted in a separate transfer pot, from which it is squirted into the mold. This can give faster cycle times, and, since no solid material is pushed around in the mold cavity itself, damage to delicate inserts (e.g., metal electrical contacts molded into a part) is minimized, mold wear is reduced and greater ease in filling intricate molds and more uniform cures are obtained. Material cured in the transfer pot and sprue is wasted, however.

19.3 EXTRUSION

Thermoplastic items with a uniform cross-sectional area are formed by extrusion. This includes many familiar items such as pipe, hose and tubing, gaskets, wire and cable insulation, sheeting, etc. Molding powder is conveyed down an electrically heated or oil-heated barrel by a rotating screw. It melts as it proceeds down the barrel and is forced through a *die* which gives it its final shape (Figure 19.4). The screw is designed to mix and compress the polymer as it progresses from solid granules to viscous melt. Vented extruders incorporate a section in which a vacuum is applied to the melt to remove volatiles such as traces of unreacted monomer, moisture, or solvent from the polymerization process.

The design of extruder screws and dies is an interesting application of the rheological principles discussed in previous chapters. The determination of the die cross section needed to produce a noncircular product cross section is still pretty much a trial-and-error process, however. Viscoelastic polymer melts swell upon emergence from the die (recovering stored elastic

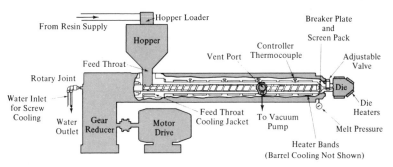

Figure 19.4.   Vented extruder (Modern Plastics Encyclopedia, Mc-Graw-Hill Book Co., 1969–70 ed.).

energy), and the degree of swell cannot be reliably predicted at the present time.

In addition to die swell, as the extrusion rate is increased, the extrudate begins to exhibit roughness and, thereafter, an irregular, severely distorted profile.   This phenomenon is known as *melt fracture.* There is not yet general agreement as to its causes, but melt fracture can be minimized by increasing die length, smoothly tapering the entrance to the die, and raising the die temperature.

Extruders are normally specified by screw diameter and length-to-diameter ratio.   Diameters range from one inch in laboratory or small production machines to a foot for machines used in the final pelletizing step of production operations.   Typical L/D ratios seem to grow each year or so, with values now in the 24/1 to 30/1 range.

The heaters on extruders are needed mainly for startup.   In steady-state operation, most or all of the heat for plasticizing the polymer is supplied by the drive motor through viscous energy dissipation.   In fact, cooling through the barrel walls and/or screw center is sometimes necessary.

Most packaging film is produced by *blow extrusion* (Figure 19.5).   A thin-walled hollow cylinder is extruded vertically upward.   Air is introduced to the interior of the cylinder, expanding it to a tube of film (less than 0.001-inch thickness).   The tube is grasped between rolls at the top, preventing the escape of air and flattening it for subsequent slitting and windup on rolls.   The expanded tube is rapidly chilled by a

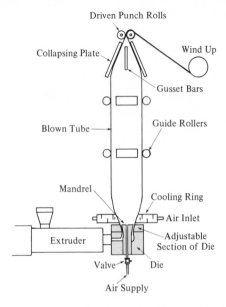

Figure 19.5.   Blow extrusion process for film production (Modern
Plastics   Encyclopedia,   McGraw-Hill   Book   Co.,
1969–70 ed.).

blast of air from a chill ring as it proceeds upward. In the case
of polyethylene, this rapid chilling produces smaller crystallites
and enhances film clarity.

19.4 BLOW MOLDING

A quick walk through a supermarket provides convincing
proof of the economic importance of plastic bottles. They're
nearly all made by blow molding (Figure 19.6). In the most
common form of this process, a hollow, cylindrical tube or
parison is extruded downward. The parison is then clamped
between halves of a cooled mold. The mold seals off the bottom of the parison and forms the threads in the neck of the
bottle. Compressed air expands the parison against the inner
mold surfaces, and, when the part has cooled sufficiently, the
mold opens, the part is ejected, and the mold is returned to
grab another parison. In a variation of this process, the

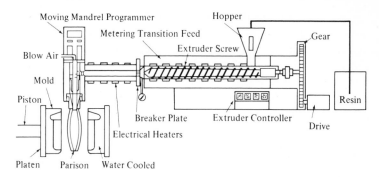

Figure 19.6. Blow molding (Modern Plastics Encyclopedia, Mc-Graw-Hill Book Co., 1969–70 ed.).

parison is formed by injection molding, usually with one end closed and the threads molded on.

The rheological properties of the parison are important. If it sags too much before being grasped by the mold, the walls of the bottle will be too thin in places. Sag is minimized by using high molecular weight compounds with high viscosities and melt elasticities. Recently, blow-molding extruders are being equipped with *parison programmers*, which vary the orifice diameter as the parison is being extruded and provide any desired variation in wall thickness in the blown product.

19.5 ROTATIONAL, FLUIDIZED-BED AND SLUSH MOLDING

Molding techniques have been developed to take advantage of the availability of finely powdered thermoplastics, mainly polyethylene. In rotational molding, the powder is introduced to a heated mold which is then rotated, bringing the powder into contact with the mold walls where it melts and takes the shape of the mold interior. The mold is then cooled and the solidified product removed. The polymer powder may also be fluidized by a stream of air. Such a fluidized bed has many of the characteristics of a liquid. When a heated form is dipped in the fluidized bed, the powder contacting it melts to form a continuous coating on it.

Similar procedures have been in use for years with plastisols (chapter 7). Liquid plastisol is poured into a heated female mold. The plastisol in contact with the mold surface solidifies, and the remainder is poured out for reuse. This *slush-molding* process is used to produce objects such as doll's heads, etc. Dipping a heated object into a liquid plastisol coats it with plasticized polymer. Vinyl-coated wire dishracks are familiar products of this process.

19.6 CALENDERING

Polymer sheet (greater than 0.001-inch thickness) may be produced either by extrusion through thin, flat dies, or by *calendering* (Figure 19.7). Basically, a calender consists of a

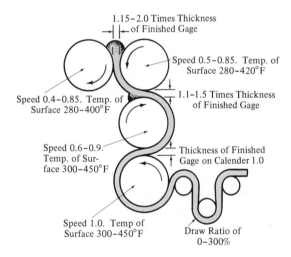

1.15–2.0 Times Thickness of Finished Gage

Speed 0.5–0.85. Temp. of Surface 280–420°F

Speed 0.4–0.85. Temp. of Surface 280–400°F

1.1–1.5 Times Thickness of Finished Gage

Speed 0.6–0.9. Temp. of Surface 300–450°F

Thickness of Finished Gage on Calender 1.0

Speed 1.0. Temp of Surface 300–450°F

Draw Ratio of 0–300%

Figure 19.7. Inverted "L" calender, illustrating process variables for the production of PVC sheet (Modern Plastics Encyclopedia, McGraw-Hill Book Co., 1969–70 ed.).

series of rotating, heated rolls, between which the polymer compound (again, most often plasticized PVC) is squeezed into sheet form, the thickness of the sheet being determined by the clearance between the rolls. Commercial calenders are very large (rolls 2 feet in diameter by 5 feet long) and represent a big capital outlay, but they are capable of tremendous production rates. In addition to forming unsupported sheeting, the polymer may be laminated to a layer of fabric between two rolls to give a supported sheeting. The final rolls may also be embossed to impart a pattern to the sheet. Shower curtains, vinyl upholstery materials (Naugahyde) and vinyl floor tile are produced by calendering.

19.7 SHEET FORMING

Thermoplastic sheet is converted to a wide variety of finished articles by processes known generically as sheet forming. Although the details vary considerably, these processes all involve heating the sheet above its softening point and forcing it to conform to a mold (Figure 19.8). In vacuum forming, for

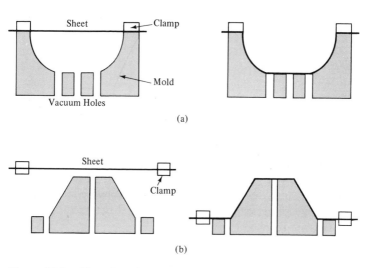

Figure 19.8.   Sheet-forming processes.   (a) vacuum forming, (b) drape forming.

example, the heat-plasticized sheet (either directly from an extruder or preheated in an oven) is drawn against the mold surface by the application of a vacuum from beneath the surface. Similarly, a positive pressure may be used to force the sheet against the mold surface, and, where very deep draws are required, mechanical assists—plug forming—are used.   In drape forming, the plasticized sheet is draped over a male mold, perhaps with a vacuum assist.

From a material standpoint, polymers used for sheet forming should have high melt strengths—i.e., high melt viscosities so that they do not draw down or thin out excessively or perhaps even tear in the forming operation. Thus, high molecular weight resins are preferred. One of the great advantages of sheet forming is that relatively inexpensive molds are required since no high pressures are involved. Epoxy molds are often used because they can be easily cast to shape from a hand-made pattern.

Among the many familiar items made by sheet forming are drinking cups, meat trays, cigarette packs, aircraft canopies, advertising displays and lighting globes. An experimental auto body has been produced by forming upper and lower halves and joining them at the centerline.

19.8 SOLUTION CASTING

Plastic sheet or film may be produced by dissolving the polymer in an appropriate solvent, spreading the viscous solution onto a polished surface, and evaporating the solvent. In the manufacture of photographic film base, the solution is spread with a knife onto a slowly rotating wheel about two feet wide and twenty feet in diameter. As the wheel revolves, heated air evaporates the solvent (which is recovered for reuse), and the dried film is stripped from the wheel before the casting point is again reached.

19.9 CASTING

The raw materials for many thermosetting polymers are available as low molecular weight liquids which are converted to solid or rubbery materials by the crosslinking reaction. These materials are easily cast to shape at atmospheric pressure in inexpensive molds. A good example of this process is the production of carved-wood furniture panels from polyester resins. Only the original need actually be carved (which is the expensive step). A mold is made by pouring a room-temperature curing silicone rubber over the original. When cured, the rubber mold is stripped from the original and used to cast the polyester copies from a mixture of a low-molecular weight unsaturated polyester and a vinyl monomer (usually styrene). A free-radical initiator causes crosslinking to the final solid object, which reproduces faithfully the detail of the carved original at a fraction of the cost. Delicate electrical components are often encapsulated or potted by casting a thermosetting liquid resin around them. Linear polymers are similarly cast from liquid monomers.

Example 1. A polymer of divinyl benzene

$$H_2C = \underset{H}{\overset{H}{C}} - \underset{}{\bigcirc} - \underset{H}{\overset{H}{C}} = CH_2$$

is to be used to make rotameter bobs (1″ long × 1/2″ diameter cylinders). What processing technique should be used?

Solution. The monomer is tetrafunctional (chapter 2); hence, it will be highly crosslinked in the polymerization reaction and therefore cannot be softened by heat. The *only* suitable technique is casting directly from the monomer, either to final shape or casting rods which can then be cut and machined to the final shape.

19.10 REINFORCED THERMOSET MOLDING

Many plastics, when used by themselves, do not possess enough mechanical strength for structural applications. When reinforced with high-modulus fibers, however, the composites have high strength-to-weight ratios and can be fabricated into a wide variety of complex shapes. Glass fibers are used extensively to reinforce thermosetting plastics, mostly polyesters and epoxies. These so-called fiberglass materials are now used for just about all boats under 40 feet in length, truck cabs, low-production volume automobile bodies, structural panels, aircraft components, etc. Such objects are fabricated by a process known as *hand layup.* The mold surface is often first sprayed with a pigmented but nonreinforced *gel coat* of the liquid resin to provide a smooth surface finish. The gel coat is followed up by successive layers of fiberglass, either in the form of woven cloth or random matting impregnated with the liquid resin, which is then cured (crosslinked) to give the finished product. This process is tremendously versatile. The molds are relatively inexpensive because no pressure is required. Objects may be selectively reinforced by adding extra layers of material where desired. Its major drawback is the expense of the hand labor involved, which makes it uneconomic for large production volumes. For automobile bodies, it becomes more expensive than stamped steel at levels above about 100,000 units per year, for example.

In many objects, stresses are not uniformly distributed, so the reinforcing fibers may be arranged to support the stress most efficiently. In fiberglass fishing rods and vaulting poles, the fibers are arranged along the long axis to resist the bending stresses applied. The process of *filament winding* extends this principle to more complex structures. Continuous filaments of the reinforcing fiber are impregnated with resin and then wound on a rotating *mandrel*. The winding pattern is designed to resist most efficiently the anticipated stress distribution. The range of the Polaris submarine ballistic missile was increased by several hundred miles by replacing the metallic rocket casing with a filament-wound reinforced plastic system. This technique is also used to produce tanks and pipe for the chemical process industries and even gun barrels by filament winding around a thin metal core which provides the necessary heat and abrasion resistance.

To provide even better structural properties, silane *coupling agents* are used to improve the bond between the resin and glass fibers. Exotic new reinforcing fibers are being developed continuously, particularly for aerospace applications where extremely high strength-to-weight ratios and good high-temperature resistance are necessary and cost is of secondary importance.

19.11 FIBER SPINNING

The polymers used for synthetic fibers are similar (and in many cases identical) to those used as plastics, but, for fibers, the processing operation must produce an essentially infinite length-to-diameter ratio. In all cases, this is accomplished by forcing the plasticized polymer through a *spinnerette*, a plate, in which a multiplicity of holes have been formed to produce the individual fibers which are then twisted together to form a thread for subsequent weaving operations. The cross-section of the spinnerette holes obviously has a lot to do with the fiber cross section, which in turn greatly influences the properties of the fiber. The three basic types of spinning operations differ mainly in the method of plasticizing and deplasticizing the polymer.

Melt spinning is basically an extrusion process. The polymer is plasticized by melting and is pumped through the spinnerette. The fibers are usually solidified by a cross-current blast of air as they proceed to the drawing rolls. The drawing step stretches the fibers, orients the molecules in the direction of stretch, and induces high degrees of crystallinity, a necessity for good fiber properties. Nylons are commonly melt spun.

In *dry spinning*, a solution of the polymer is forced through the spinnerette. As the fibers proceed downward to the drawing rolls, a countercurrent stream of warm air evaporates the solvent. In this process, the cross-section of the fiber is determined not only by the shape of the spinnerette holes but also by the complex nature of the diffusion-controlled solvent evaporation process because there is considerable shrinkage as the solvent evaporates. The acrylic fibers (Orlon, Creslan, Acrilan, etc.), mainly polyacrylonitrile, are produced by dry spinning.

Wet spinning is similar to dry spinning in that a polymer solution is forced through the spinnerette. Here, however, the solution strands pass directly into a liquid bath. The liquid might be a nonsolvent for the polymer, precipitating it from solution as the solvent diffuses outward into the nonsolvent. The bath might also contain a substance which precipitates a polymer fiber by chemically reacting with the dissolved polymer. Here again, the process as well as the spinnerette influences the fiber cross section. Rayon (regenerated cellulose) is a common example of a wet-spun fiber.

Example 2. Suggest processing techniques for the manufacture of

1. 100,000 ft of plasticized PVC garden hose,
2. 50,000 polystyrene pocket combs,

3. 100,000 polyethylene detergent bottles,

4. 5,000 phenolic (phenol-formaldehyde) TV knobs,

5. 6 souvenir paperweights of polymethyl methacrylate containing an old coin,

6. a strip of chlorinated rubber (a linear, amorphous polymer) and white pigment (60:40), roughly 0.001 inches thick, 4 inches wide, running 20 miles down the center of a highway, and

7. 1,000,000 polystyrene meat trays, 0.005 inches thick.

Solution

1. Extrusion.

2. Injection molding.

3. Blow molding.

4. Compression, transfer, or thermoset injection molding.

5. Although these objects could be injection molded, the small number of articles required would make it uneconomic. They can easily be *cast* from the monomer.

6. Dissolve rubber in solvent, add pigment and spray on highway. Evaporation of solvent leaves the desired strip.

7. Extrude sheet, then vacuum form trays.

GENERAL REFERENCES

1. Bernhardt, E. C., ed. *Processing of thermoplastic materials*, Reinhold Publishing Corp., New York, 1959.

2. McKelvey, J. M. *Polymer Processing*, John Wiley & Sons, Inc., New York, 1962.

3. Miles, D. C., and J. H. Briston, *Polymer Technology*, Part III. Chemical Publishing Co. Inc., New York, 1965.

4. *Modern Plastics Encyclopedia.* McGraw-Hill Book Co., New York, (annually).

5. Pearson, J. R. A. *Mechanical Principles of Polymer Melt Processing*, Pergamon Press, Ltd., Oxford, 1961.

6. Rodriguez, F. *Principles of Polymer Systems*, ch. 12. McGraw-Hill Book Co., New York, 1970.

7. Schildknecht, C. E., ed. *Polymer Processes*, ch. XVI and XVIII. John Wiley & Sons, Inc., New York, 1956.

20
Plastics

20.1 INTRODUCTION

As was mentioned in the Introduction, there are five major applications for polymers: plastics, rubbers, synthetic fibers, surface finishes and adhesives. Previous sections have dealt with the properties of the polymers themselves. While these properties are undoubtedly the most important in determining the ultimate application, polymers are rarely used in a chemically pure form so, in a discussion of the technology of polymers, it is necessary to mention, in addition to the properties of the polymers required for a given application, the nature and reasons for use of the many other materials often associated with the polymers. The following brief chapters will do this in turn for the five major applications.

Plastics are normally thought of as being polymer compounds possessing a degree of structural rigidity—in terms of the usual stress-strain test, a modulus on the order of 10^9 dynes/cm^2 or greater. The *molecular requirements* for a polymer to be used in a plastic compound are (a) if linear or branched, the polymer must be below its glass transition temperature (if amorphous) and/or below its crystalline melting point (if crystallizable) at use temperature, or (b) it must be crosslinked sufficiently to restrict molecular response essentially to straining of bond angles and lengths (e.g., ebonite or hard rubber).

20.2 MECHANICAL PROPERTIES OF PLASTICS

The engineering properties of commercial plastics vary considerably within the broad definition given above. Figure 20.1

251

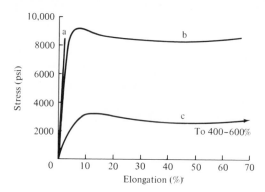

Figure 20.1.   Typical stress-strain curves for plastics.   (a) rigid and brittle, (b) rigid and tough, (c) flexible and tough.

sketches some representative stress-strain curves for three types of plastics. As discussed previously, for any given material, the quantitative nature of the curves depends markedly on the rate of strain and the temperature. In general, faster straining and lower temperatures lead to higher moduli (slopes) and smaller ultimate elongations.

Plastics with stress-strain curves of the type (a) in Figure 20.1 are *rigid* and *brittle*. The former term refers to the high initial modulus. The latter refers to the area under the stress-strain curve which represents the energy per unit volume required to cause failure. These materials usually fail by catastrophic crack propagation at strains on the order of 2 percent. Since *hardness* correlates well with tensile modulus, it is another valuable property of this type of plastic. Examples of this class are polystyrene, polymethyl methyacrylate and most thermosets. Curve (b) represents *rigid* and *tough* materials, sometimes known as *engineering thermoplastics*. In addition to high modulus, tensile strength and hardness, these materials undergo *ductile deformation* or drawing beyond the yield point, which is evidence of considerable molecular orientation before failure. The latter confers toughness or *impact resistance*—i.e., the ability to withstand shock loading without brittle failure. Examples of this class of materials are polycarbonates, cellulose esters, nylons, and polypropylene. Curve (c) represents *flexible* and *tough* plastics, as typified by low- and medium-density polyethylenes. Here, the ductile deformation leading to very high ultimate elongations results from the conversion of low-crystallinity material to high-crystallinity material. The tensile samples 'neck down,' with the density, percent crystallinity and modulus of the material in the neck being appreciably greater than those of the parent material. Despite their good toughness, these materials are limited in their structural applications by their low moduli and tensile strengths.

20.3 CONTENTS OF PLASTIC COMPOUNDS

In addition to the polymer itself, which is seldom used alone, plastics usually contain at least small amounts of one or more of the following additives.

1. Reinforcing Agents The function of these is to enhance the structural properties of the compound, in particular properties such as modulus and the retention of modulus at higher temperatures. The use of long glass fibers to reinforce epoxy and polyester thermosets was mentioned in the previous chapter. A relatively recent and increasingly important example of reinforcement is the use of short (1/8- to 1/2-inch) glass fibers to reinforce common thermoplastic polymers. Addition of up to 40 weight percent short glass fibers to polystyrene, nylon, polypropylene, etc., provides a relatively inexpensive way of greatly improving the structural strength of the plastic and maintaining its strength to higher temperatures, as illustrated in Table 1. (see page 254.) As might be expected, the addition of the fibers increases the processing difficulty somewhat, but the commercially available compounds are routinely injection molded.

Another example of reinforcement is the use of dispersed rubber particles (on the order of a micron in diameter) to increase the *impact strength* of a normally brittle glassy polymer. Figure 2 illustrates the effect of adding increasing levels of a rubber to polystyrene in a high rate-of-strain (to simulate shock loading) tensile test. The addition of the rubber particles decreases the modulus and ultimate tensile strength, but it introduces considerable ductile deformation, thereby increasing the area under the stress-strain curve. For many applications, the improvement in impact strength considerably out-

TABLE 1   Some Properties of Short Glass Fiber Reinforced
Thermoplastics (Fiberfil®)

	Nylon 66		Polystyrene		Polycarbonate	
	unreinf.	reinf.	unreinf.	reinf.	unreinf.	reinf.
Tensile						
Strength, psi	11,800	20,000	8,500	14,000	9,000	20,000
Elongation, %	60	1.5	2.0	1.1	60–100	1.7
Tensile						
Modulus, psi	400,000	1,000,000	400,000	1,100,000	320,000	1,300,000
Flexural						
Strength, psi	11,500	28,000	11,000	17,500	12,000	26,000
Izod Impact						
(¼ × ½ in.						
bar) ft-lb/in.	0.9	2.5	0.3	2.9	2.0	4.0
Heat-distortion						
point (264						
psi) °F	150	498	190	220	280	300
Rockwell						
Hardness	M79	M100	M70	M95	M70	M95

weighs the slight decreases in modulus and tensile strength produced by the addition of rubber. A more complete discussion of these two-phase polymer systems is available elsewhere (*1*).

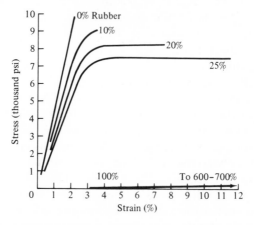

Figure 20.2.   The influence of discrete, micron-sized rubber particles on the high-speed (133 in/in min) stress-strain curves of polystyrene (*1*).

2. Fillers   An important function of many fillers in plastics is simply to cheapen the compound; i.e., they are low-cost materials with only secondary beneficial (and sometimes detrimental) effects on properties. Wood flour (fine sawdust) in phenolic or other thermosets is a good example. Recently, a series of polyester resins filled with water droplets (water-extended polyesters—WEP) has been introduced for decorative applications (statuary, carved furniture panels, etc.) where great structural strength is not necessary. It's hard to imagine a cheaper filler, and the water is said to impart wood-like machining properties to the compound. Sometimes fillers may improve a certain property. For example, the addition of mica or asbestos to thermosets increases the compound's heat resistance. They also have some mechanical reinforcing abilities and might be included in the previous category.

3. Stabilizers   These compounds serve to prevent or inhibit chemical degradation of the polymer. In the case of polyvinyl chloride, for example, the major product of thermal degradation is HCl, which catalyzes further degradation. Compounds which react with the HCl to form stable products, such as metal oxides, are used as stabilizers. Oxidative degradation of polymers is thought to take place by a free-radical mechanism involving crosslinking and/or chain scission initiated by free radicals from peroxides formed in the initial oxidation step. Similarly, these reactions can be initiated by free radicals produced by ultraviolet radiation. Stabilizers against such reactions are generally quinone-type organics which are effective free-radical scavengers.

4. Pigments   Plastic compounds may be colored by the addition of small amounts of finely divided solids, giving an opaque material, or by organics which dissolve in the polymer to produce a transparent compound (assuming the polymer is transparent to begin with). Some thermoplastics, although transparent, have a slight yellow tinge caused by selective absorption of light toward the blue end of the spectrum. This can largely be cancelled out by the addition of a blue pigment which reduces the transmission of yellowish wavelengths. This technique unavoidably results in a slight lowering of the total light transmission. Carbon black often performs the dual function of pigment and stabilizer in plastics.

It prevents degradation by absorbing and preventing the penetration of ultraviolet light beyond the surface of the article.

5. Plasticizers   The addition of a relatively low molecular weight organic plasticizer to a normally glassy polymer will progressively reduce its modulus by bringing the compound's glass transition temperature down closer to use temperature. The level at which the modulus is reduced to the point where the compound is considered an elastomer rather than a plastic is somewhat arbitrary, but plastics sometimes contain small amounts of a plasticizer to reduce brittleness.

6. Lubricants   Certain low molecular weight organics which are relatively insoluble in the polymer will sweat out or migrate to the surface of the compound and form a slippery coating during processing operations. These lubricants produce smoother extrudates and molded articles, and they minimize sticking in the mold. Stearic acid and its metal salts are common lubricants.

7. Curing Agents   These are chemicals whose function is to produce a crosslinked, thermosetting plastic from an initially linear or branched polymer. A vinyl monomer, such as styrene, a free-radical initiator and sometimes a *promoter* (e.g., cobalt naphthenate—to speed up the reaction) are dissolved in a low molecular weight unsaturated polyester resin and crosslink it by an ordinary addition mechanism involving the double bonds in the polyester.

The polymer for a *three-step* thermosetting compound is deliberately produced with a stoichiometric shortage of one of the reagents to give a highly branched but not yet crosslinked structure which is later cured in the mold with a curing agent. An A-stage phenolic polymer is produced by reacting phenol and formaldehyde in about a 1.25/1 mole ratio (a molal excess of formaldehyde is required for crosslinking). This still-thermoplastic product is compounded with fillers, pigments, reinforcing agents, etc., and a curing agent, hexamethylenetetramine, to give a B-stage resin. The curing agent decomposes in the presence of moisture (a product of the condensation reaction) upon heating in the mold, giving the additional formaldehyde, required for crosslinking, and ammonia, which acts as a basic catalyst for the reaction.

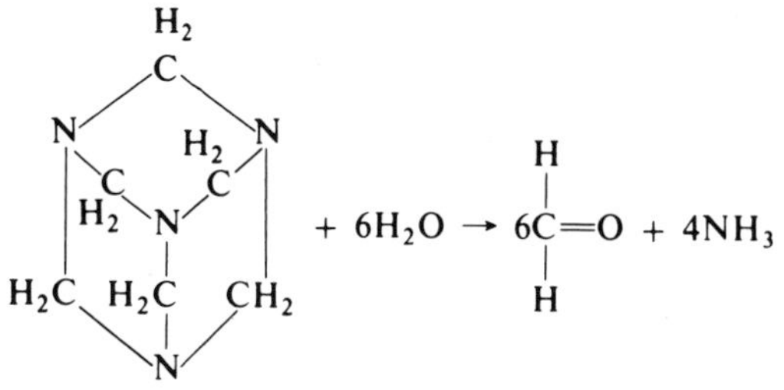

$$\text{hexamethylenetetramine} + 6H_2O \rightarrow 6\overset{\displaystyle H}{\underset{\displaystyle H}{C}}{=}O + 4NH_3$$

hexamethylenetetramine   water   formaldehyde   ammonia

Two-step resins are made by using the final desired reactant ratio but stopping the reaction short of crosslinking. The reaction is completed in the mold without the necessity of a curing agent.

Free-radical initiators (usually organic peroxides) are used as curing agents for saturated thermoplastic polymers. The curing agent must be compounded with the polymer at temperatures low enough to prevent appreciable decomposition to free radicals. The compound is then heated in the mold, producing free radicals which abstract protons from the polymer, leaving unshared electrons on the chains which combine to form crosslinks.

$$R{:}R \quad \rightarrow \quad 2R\cdot$$

initiator   heat   free radical
(curing agent)

$$R\cdot + \text{~}C{-}C{-}C\text{~} \rightarrow \text{~}C{-}C{-}C\text{~} + R{:}H$$

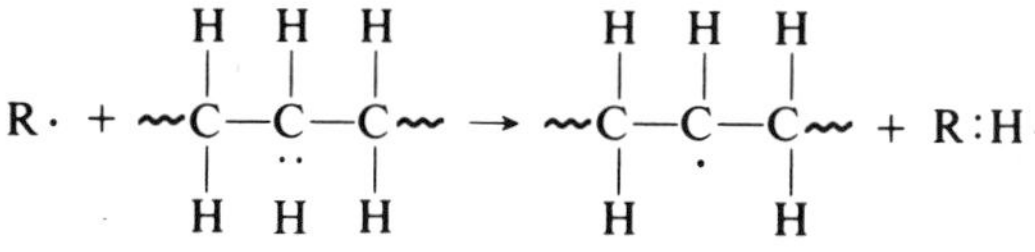

polymer chain

$$2\ \text{~}C{-}C{-}C\text{~} \rightarrow \text{crosslinked chains}$$

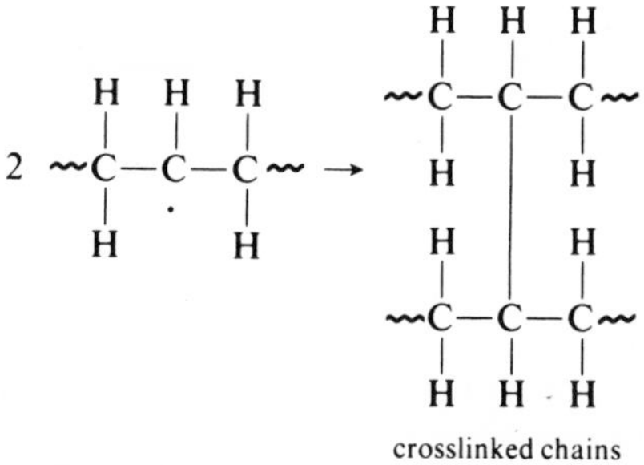

crosslinked chains

Sometimes, sulfur or multifunctional vinyl monomers are added to improve crosslinking efficiency by helping to bridge the gap (i.e., produce longer crosslinks) between the chains. In this manner, polyethylene is crosslinked with dicumyl peroxide, converting it from a thermoplastic to a thermoset with much greater heat resistance and resistance to stress cracking and abrasion. It can also tolerate much higher levels of carbon black filler without becoming excessively brittle.

8. Blowing Agents   Foamed plastics must contain a material which generates a gas to produce the foaming. These blowing agents may either generate gas through a chemical reaction or simply vaporize upon heating. Polyurethane foams are produced by the reaction of excess isocyanate groups with water, generating CO_2,

$$R-N=C=O + H_2O \rightarrow RNH_2 + CO_2$$
$$\text{isocyanate} \qquad\qquad \text{amine}$$

or, alternately, by using an inert but volatile fluorocarbon blowing agent.

Polystyrene foams are produced by dissolving an inert but volatile hydrocarbon liquid such as heptane in the polymer. In the manufacture of foam molding beads, the blowing agent is added to the monomer in a suspension polymerization. When the resulting beads are placed in a mold and heated (usually with steam), the hydrocarbon volatilizes, expanding the beads against each other and the mold walls. These beads are used in many familiar applications such as drinking cups, picnic coolers and packaging supports.

REFERENCE

1. Rosen, S. L. Two-phase polymer systems. *Polymer Eng. and Sci.*, **7**, no. 2, p. 115, 1967.

21
Rubbers

21.1 INTRODUCTION

A rubber is generally defined as a material which can be stretched to at least twice its original length and which will retract rapidly and forcibly to substantially its original dimensions on release of the force. An elastomer is a rubber-like material from the standpoint of modulus but which has limited extensibility and incomplete retraction. The most common example is highly plasticized polyvinyl chloride.

From a molecular standpoint, a rubber must be (a) a high polymer, since rubber elasticity is due mainly to the uncoiling and coiling of long chains (chapter 14). In order that the molecules can be able to coil and uncoil freely, (b) the polymer must be above its glass transition temperature at use temperature. Furthermore, (c) the polymer must be amorphous in its unstretched state, as crystallinity would hinder the molecular motion necessary for rubber elasticity. Until very recently, an additional requirement was that (d) the polymer be crosslinked. If this were not so, the chains would slip past one another under stress (viscous flow), and recovery would be incomplete.

21.2 THERMOPLASTIC ELASTOMERS

The introduction of so-called thermoplastic elastomers (actually rubbers) has gotten around the requirement for cross-linking in the strictest sense of the term—covalent bonding between chains. These polymers are styrene—butadiene—styrene (SBS) block copolymers, produced in solution by anionic polymerization (chapter 11). Since most polymer pairs are mutually incompatible (insoluble), due largely to the very

259

low entropies of solution (chapter 7), the polystyrene chains ends aggregate together in microscopic domains. These polystyrene domains, being below their glass transition temperature (100C) at normal use temperatures, are rigid, and they act to tie together the long, flexible polybutadiene segments (above their T_g) as do ordinary crosslinks. Unlike the usual covalent crosslinks, however, the polystyrene domains melt above the T_g of polystyrene, and the polymer behaves as a true thermoplastic.

21.3 CONTENTS OF RUBBER COMPOUNDS

Natural rubber, and with a few exceptions the many synthetic rubber polymers commercially available, are unsaturated; i.e., they contain double bonds which provide sites for the vulcanization (crosslinking) reaction. The polymers are mostly linear or branched, but many contain also (either intentionally or unintentionally) quantities of gel (crosslinked particles) before vulcanization. This can have a profound effect on processing properties. Also included in rubber compounds are the following.

1. Reinforcing Agents Any rubber compound which is to develop a reasonable tensile strength, abrasion and tear resistance will contain up to 50 phr (parts per hundred parts rubber by weight) or more of a reinforcing agent, nearly always a carbon black. The use of carbon black in rubbers is quite different than in plastics where it is strictly a pigment and limited to much lower loadings. Carbon blacks are not simply carbon. Basic blacks have hydroxyl groups at the particle surfaces, and acid blacks have carboxyllic acid functionality. It has been shown that the rubber polymer forms strong secondary and primary covalent bonds with the carbon black, which accounts for its reinforcing ability. There is a wide variety of carbon blacks available. In addition to chemical functionality, they differ in such factors as particle size, degree of aggregation, surface area, etc., and different types of rubber polymers require particular kinds of black for optimum reinforcement. Silicone rubbers are sometimes reinforced with finely divided silica (SiO_2).

Natural rubber (cis-1,4 polyisoprene) and its synthetic counterpart and butyl rubber are among the few rubber polymers which can develop reasonable mechanical strength without reinforcement. This is because they crystallize with molecular orientation at high elongations, and the crystallites function as a reinforcing agent, causing the sharp upturn in the stress-strain curves at high elongations shown in Figure 21-1 (*1*).

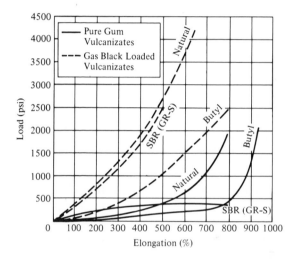

Figure 21.1.   Stress-strain curves for filled and unfilled rubber vulcanizates (*1*). (From Principles of High-Polymer Theory and Practice by A. X. Schmidt and C. A. Marlies. Copyright 1948 by the McGraw-Hill Book Company, Inc. Used with permission of McGraw-Hill Book Company.)

2. Fillers   As with plastics, the function of fillers in rubber is mainly to reduce the cost of the compound. The most common rubber fillers are finely divided inorganics such as $CaCO_3$. The addition of such high-modulus fillers raises the modulus (stiffens the compound), of course, and too much will cause a loss of rubbery properties, all other things being equal. Carbon blacks sometimes perform as fillers as well as reinforcing agents in rubber compounds.

3. Extending Oils   Hydrocarbon oils are often used in rubber compounds. Their function is twofold. First, they plasticize the polymer, making it softer and easier to process. This is particularly important with very high molecular weight polymers. Second, since they are considerably cheaper than the rubber polymer, they act like fillers in reducing the cost of the compound. Extending oils are available in various degrees of aromaticity, and the properties of the compound will depend on the type of oil in relation to the polymer as well as oil level.

Carbon black and extending oils have opposite effects on the modulus of a rubber compound. One of the newer tricks for producing low-cost compounds from synthetic polymers is to polymerize to higher-than-normal molecular weight. The modulus is cut down by the addition of an extending oil and then brought back up to the desired level by the addition of large amounts of black. By projecting figures showing the increasing levels of black and oil in rubber compounds, it appears that, by about 1980, there won't be any polymer left. Seriously, there is a limit to this technique, as other properties soon are degraded excessively.

4. Vulcanizing or Curing Systems   The sole function of these is to crosslink the polymer. The most common curing systems are based on sulfur. While sulfur alone will cure unsaturated rubbers with heating, the process is slow and inefficient in its use of the sulfur. The mechanisms of sulfur curing are not well understood but are thought to include (among other things) the formation of sulfide or disulfide links between chains and the abstraction of protons from adjacent chains to form H_2S, with the chains crosslinking at the remaining unshared electrons.

To speed up the vulcanization process, *accelerators* are generally used. These are usually complex sulfur-containing organic compounds, often of proprietary composition. *Promotors* or *activators* are used to improve the cure still further. Zinc oxide is a common example, particularly in conjunction with stearic acid.

Nonsulfur cures are used occasionally. Free-radical initiators will provide crosslinks, as discussed in the previous chapter, and zinc oxide will crosslink polymers containing chlorine.

5. Antioxidants or Stabilizers These are particularly important with natural rubber and the many synthetic rubbers which contain a major proportion of butadiene. These polymers are highly unsaturated, and the double bonds are extremely susceptible to attack by oxygen and ozone, resulting in embrittlement, cracking and general degradation. Since the degradation reactions take place by free-radical mechanisms, antioxidants are compounds which scavenge free radicals.

6. Pigments Where great mechanical strength is not required, and thus carbon black is not used, rubbers can be colored with pigments as can plastics.

A typical tire-tread formulation is shown in Table 1. Notice that the final compound is less than 60 percent polymer.

TABLE 1   Tire-Tread Formulation (2)

Ingredient	Parts by Weight
GR-S 1000 (75/25 butadiene/styrene emulsion copolymer)	100
HAF black	50
Zinc oxide ⎫ (promotors)	5
Stearic acid ⎭	3
Sulfur	2
Santocure (accelerator)	0.75
Circosol 2XH (extending oil)	10

21.4 COMPOUNDING

The ingredients above must be *compounded* with the rubber polymer to produce the final rubber compound for molding, extrusion, etc. This is usually done in *two-roll* mills or *Banbury* mills. The two-roll mill shears the material in a gap between two counter-rotating rolls. The rolls rotate at different rates, producing tremendous shear rates within the gap. The Banbury accomplishes a similar thing by the rotation of meshing blades within a closed chamber. The rolls or the Banbury chamber may be steam, oil or electrically heated, or perhaps water cooled. Both devices are driven by relatively large electric motors, and they put a lot of energy per unit time and volume into the polymer. This energy input serves two purposes.

First, it breaks down the polymer and reduces its nerve to the point where it can be easily compounded and processed. *Nerve* is a term used to describe the difficult-to-handle highly elastic response caused by too high a molecular weight. This mastication step mechanically degrades the polymer, lowering its molecular weight and therefore increasing viscous response. Premastication is particularly important with natural rubber, since no control can be exercised over its molecular weight in the polymerization step. Second, the high energy input breaks up and disperses the compounding ingredients evenly throughout the polymer.

REFERENCES

1. Schmidt, A. X. and C. A. Marlies. *Principles of high polymer theory and practice.* McGraw-Hill, New York, p. 537, 1948.

2. Garvin, G. S. *Polymer Processes*, chapter XVI. (C. E. Schildknecht, ed.). John Wiley & Sons, Inc. New York, 1956.

22
Synthetic Fibers

22.1 INTRODUCTION

While many of the polymers used for synthetic fibers are identical to those in plastics, the two industries grew up separately with completely different terminologies, testing procedures, etc. Many of the requirements for fabrics are stated in nonquantitative terms such as 'hand' and 'drape' which are difficult to relate to normal physical property measurements but which can be critical from the standpoint of consumer acceptance and, therefore, commercial success of a fiber.

A fiber is often defined as an object with a length-to-diameter ratio of at least 100. Synthetic fibers are spun in the form of continuous *filaments*, but they may be chopped up to much shorter *staple* which is then twisted into thread before weaving. Natural fibers, with the exception of silk, are initially in staple form. The thickness of a fiber is most commonly expressed in terms of *denier*, which is the weight in grams of a 9000-meter length of the fiber. Stresses and tensile strengths are reported in terms of *tenacity*, with units of grams/denier.

22.2 FIBER PROCESSING

Synthetic fibers are pretty much unoriented as they emerge from the spinning operation. To develop the tensile strengths and moduli necessary for textile fibers, they must be drawn (stretched) to orient the molecules along the fiber axis and develop high degrees of crystallinity. All successful fiber-forming polymers are crystallizable, and so, from a molecular standpoint, the polymer must have polar groups, between which strong hydrogen bonding holds the chains in a crystal lattice

265

(e.g., polyacrylonitrile, nylons), or be sufficiently regular to pack closely in a lattice held together by dispersion forces (e.g., isotactic polypropylene).

22.3 DYING

The dying of fibers is a complex art in itself. A successful dye must either form strong secondary bonds to polar groups on the polymer or react to form covalent bonds with functional groups on the polymer. Furthermore, since the fibers are dyed after spinning, the dye must penetrate the fiber and diffuse into it from the dye bath. The size of the dye molecule is such that it can't penetrate crystalline areas of the polymer, so it is mainly the amorphous regions which are dyed. This often conflicts with the requirement of high crystallinity. The chains of polyacrylonitrile, for example, while possessing the necessary polar sites for dye attachment in abundance, are so strongly bound to each other that it's difficult for the dye to penetrate. For this reason, acrylic fibers usually contain minor amounts of plasticizing comonomers to enhance dye penetration. Nonpolar, nonreactive fibers such as polypropylene, on the other hand, have no sites to which the dye can bond even if it could penetrate. This was long a problem with polypropylene fibers, and it was overcome by incorporating a finely divided solid pigment in the polymer before melt spinning. Copolymers of propylene and a monomer with dye-accepting sites are now available.

22.4 EFFECTS OF HEAT AND MOISTURE

The polarity of the polymer also directly influences the degree of water absorption of the polymer. Other things being equal, the more polar the polymer, the higher its equilibrium moisture content under any given conditions of humidity. As with dyes, however, moisture content is reduced by strong interchain bonding. The moisture content exerts a strong influence on the feel and comfort of fibers. Hydrophobic fibers tend to have a clammy feel in clothing and can build up static electricity charges. Perhaps the most important effect of moisture on polar polymers is its action as a plasticizer. Since fiber-forming polymers are linear, heat is also a plasticizer. This explains why suits wrinkle on hot, humid days and why the wrinkles can be removed by steam pressing. The increasingly popular wash-and-wear and permanent-press fabrics are produced by operations which crosslink the fibers by reacting with functional groups on the chains, such as the hydroxyls on cellulose. The more hydrophobic polymers are inherently more wrinkle resistant because they are not plasticized by water. Wash-and-wear shirts, therefore, usually are made of blends of polyethylene terephthalate and cotton, about 65/35 percent.

23
Surface Finishes

Nearly all surface finishes and coatings with the exception of ceramic types for high-temperature applications are based on a polymer film of some sort. They account for the use of a lot of polymer, but determining just how much and which polymers is not easy because most formulations are proprietary and production figures do not always separate polymer and nonpolymer components. Basically, there are five varieties of surface finishes.

23.1 LACQUERS

A lacquer consists of a polymer solution to which a pigment has been added. The film is formed simply by evaporation of the solvent, leaving the pigment trapped in the polymer film. Since no chemical change occurs in the polymer, it retains its original solubility characteristics. Hence, the major drawback to lacquers is their poor resistance to organic solvents. Much of the technology of lacquers involves the development of polymers which form tougher, more adhesive, and more stable films and the choice of solvent systems which provide the optimum viscosity for application. The volatility of the solvent system is important, too. If it is too high, it will evaporate before the film has had a chance to level, leaving brush marks or an 'orange-peel' or rough surface when sprayed. If too low, the coating will sag excessively after application.

A wide variety of polymers is used for lacquers. Newer systems favor acrylic polymers for their superior chemical stability. Also under development are systems in which the polymer film is crosslinked after application by gamma or electron radiation. The radiation energy knocks off protons, leaving unshared electrons which form crosslinks. This toughens the film and overcomes the major drawback of lacquers by making the film insoluble.

The term *spirit varnish* is an old and imprecise name for a lacquer in which the solvent is alcohol (e.g., shellac). Since the polymers are soluble in a highly polar solvent, spirit varnishes have rather poor water resistance.

23.2 OIL PAINTS

These popular and widely used finishes consist of a suspension of pigment in a *drying oil* (e.g., linseed oil, a low molecular weight, unsaturated oil). The film is formed by a reaction involving atmospheric oxygen which polymerizes and crosslinks the drying oil through its double bonds. Sometimes an inert solvent is added to control viscosity, and catalysts (e.g., cobalt naphthenate) are used to promote the crosslinking reaction. Oil paints, once cured, are no longer soluble, although they may be softened considerably by appropriate solvents (as used in paint removers).

23.3 OIL VARNISH OR VARNISH

These coatings consist of a polymer, either natural or synthetic, dissolved in a drying oil, with perhaps an inert solvent to control viscosity and a catalyst to promote the crosslinking reaction with oxygen. When cured, they produce a clear, tough, solvent-resistant film.

23.4 ENAMEL

An enamel is a pigmented oil varnish. It is much like an oil paint except that the added polymer, replacing some of the drying oil, provides a tougher, glossier film.

23.5 LATEX PAINTS

These versatile finishes are well on their way to replacing oil paints and enamels for home use, as a result of their quick-drying, low-odor, and water-cleanup properties. Basically, they are polymer latexes, produced by emulsion polymerization (chapter 13) to which pigments and rheological-control agents have been added.

The film is formed by coalescence of the polymer particles upon evaporation of the water. The polymer itself is not water soluble. Although they are sometimes termed water-soluble paints, this is a serious misnomer as is well known to anyone who has tried to clean a brush after the water has evaporated. As long as the individual latex particles have not coalesced, they are water-*dispersable*.

In order that the particles coalesce to form a film when the water evaporates, the polymer must be deformable under the action of surface-tension forces. Thus, polymers for latex paints must be near or above their glass transition temperatures at use temperature. Early latex paints were based on styrene-butadiene copolymers or vinyl acetate. These are being supplanted by paints based on acrylic (ethyl and other acrylates) latexes because of the acrylics' superior chemical stability and, therefore, resistance to color change and degradation.

Much work has gone into adjusting the formulations of latex paints to allow high pigment contents and thick films for one-coat coverage, together with reasonable ease of application. The main rheological property desired is *thixotropy* (chapter 15), which gives a high viscosity to prevent the settling of pigments and sagging and dripping of the applied film together with a lower viscosity under the shearing of brushing or rolling for easy application. These properties are achieved by the addition of finely divided inorganics and water-soluble polymers.

24
Adhesives

Although not often thought of as an application for polymers, adhesives are extremely important technologically and are becoming more so as new formulations and applications are continuously being developed. It is difficult to obtain figures on the use of polymers as adhesives, because many of the polymers used are the same as those used in other applications. Powers (*1*) has listed the advantages of adhesives in joining components (adherends): (a) lower stress concentration in the adherends may be obtained; (b) materials may be bonded together for which other methods of assembly are impractical; (c) adhesives may act as seals against moisture, gases or oils in addition to securing materials in place; and (d) hand labor may be reduced and other economies may be effected.

In terms of weight, it doesn't take much adhesive to join much larger adherends. Hence, it is not surprising that many of the newer high-performance adhesives have had their origins in the aerospace industry.

The science of adhesion is not yet well developed. Nevertheless, some important generalizations can be drawn. Adhesion results from (a) mechanical bonding between the adhesive and adherend and (b) chemical forces—either primary covalent bonds or polar forces—between the two. The latter are thought to be the more important, and this explains why nonpolar polymeric substrates (e.g., polyethylene and polytetrafluoroethylene) are very difficult to adhesive bond—they must first be chemically treated to introduce polar sites on the surface.

To a good approximation, the properties of the adhesive polymer determine the properties of the adhesive joint; i.e., the bond can be no stronger than the glue line. Brittle polymers give brittle joints, polymers with high shear strengths give bonds of high shear strength, heat-resistant polymers produce bonds with good heat resistance, etc.

In order that a successful joint be formed, the adhesive must contact the surface intimately; i.e., it must *flow* into nooks and crannies and it must *wet* the surfaces to be joined. To provide the necessary mechanical strength, it must *harden* once it has made good contact with the surfaces. There are five general categories of adhesives which accomplish these objectives in different ways.

24.1 SOLVENT-BASED ADHESIVES

Here the adhesive polymer is made to flow by dissolving it in an appropriate solvent. They solidify by evaporation of the solvent. Thus, the polymers used must be linear or branched to allow solution, and the joints formed will not be resistant to solvents of the type used initially to dissolve the polymer. To get a good bond, it often helps if the solvent attacks the adherend also. In fact, solvent alone is often used to *solvent-weld* the polymers, dissolving some of the adherend to form an adhesive on application. One of the drawbacks to solvent-based adhesives is the shrinkage which results when the solvent evaporates. This can set up stresses which weaken the joint. Examples of this type of adhesive are the familiar model-airplane cement and rubber cement.

24.2 LATEX ADHESIVES

These materials are based on polymer latexes made by emulsion polymerization. They flow easily while the continuous water phase is present, and they dry by evaporation of the water, leaving behind a layer of polymer. In order that the polymer particles coalesce to form a continuous joint and be able to flow to contact the adherend surfaces, the polymers used must be above their glass transition temperature at use temperature. These requirements are identical to those for latex paints, so it is not surprising that the same polymers are used in both applications (e.g., styrene-butadiene copolymers and vinyl acetate). Nitrile rubber (acrylonitrile-butadiene copolymer) is used for increased polarity. A familiar example of a latex adhesive is white glue, basically a high-solids polyvinyl acetate latex. Latex adhesives are used extensively for bonding pile to backing in carpets.

24.3 PRESSURE-SENSITIVE ADHESIVES

These are really highly viscous polymer melts at room temperature, so the polymers used must be above their glass transitions. They are caused to flow and contact the adherends by applied pressure, and, when the pressure is released, the viscosity is high enough to withstand the stresses produced by the adherends, which obviously cannot be very great. The many varieties of pressure-sensitive tape are faced with this type of adhesive.

24.4 HOT-MELT ADHESIVES

Thermoplastics often form good adhesives simply by being melted to cause flow and then solidifying on cooling after contacting the surfaces under moderate pressure. Nylons are used frequently as hot-melt adhesives. Electric 'glue guns' have recently been introduced to the consumer market which operate on this principle.

24.5 REACTIVE ADHESIVES

These compounds are either monomeric or low-molecular weight polymers which solidify by a polymerization and/or crosslinking reaction after application. They are particularly effective if they can react chemically with the adherend. Epoxies, phenolics, cyanoacrylates and silicone rubbers are important examples of this category. They can develop tremendous bond strengths, and they have good solvent resistance and good (for polymers, anyhow) high-temperature properties. They are not always initially in liquid form. For example, a B-stage (chapter 20) phenolic resin may be used to impregnate tissue paper. These dry sheets are placed between layers of wood in a heated press where the resin first melts and then cures, bonding the wood to form plywood.

REFERENCE

1. Powers, W. J. *Polymer Processes*, ch. XII (C. E. Schild-knecht, ed.). John Wiley & Sons, Inc., New York, 1956.

Selected Readings

Alfrey, T. and E. F. Gurney. *Organic Polymers.* Prentice-Hall Inc., Englewood Cliffs, N.J. 1967.

Baer, E. ed. *Engineering Design for Plastics.* Reinhold Publishing Corp., New York, 1964.

Billmeyer, F. *Textbook of Polymer Science.* John Wiley & Sons, Inc., New York, 1962.

Brandrup, J. and E. H. Immergut. *Polymer Handbook.* John Wiley & Sons, Inc., New York, 1966.

Brydson, J. A. *Plastics Materials.* Van Nostrand Co., Inc., Princeton, N.J., 1966.

Coleman, B. D., H. Markovitz and W. Noll. *Viscometric Flows of Non-Newtonian Fluids.* Springer-Verlag Inc., New York, 1966.

Ferry, J. D. *Viscoelastic Properties of Polymers.* John Wiley & Sons, Inc., New York, 1961.

Flory, P. J. *Principles of Polymer Chemistry.* Cornell University Press, Ithaca, New York, 1953.

Golding, B. *Polymers and Resins.* Van Nostrand Co., Inc., Princeton, N.J., 1959.

Gorden, M. *High Polymers.* Addison-Wesley Pub. Co. Inc., New York, 1963.

Lenz, R. W. *Organic Chemistry of Synthetic High Polymers.* John Wiley & Sons, Inc., New York, 1967.

McKelvey, J. M. *Polymer Processing.* John Wiley & Sons, Inc., New York, 1962.

Meares, P. *Polymers: Structure and Bulk Properties.* Van Nostrand Co. Inc., Princeton, N.J., 1965.

Miles, D. C. and J. H. Briston. *Polymer Technology.* Chemical Publishing Co., New York, 1965.

Miller, M. L. *The Structure of Polymers*. Reinhold Publishing Corp., New York, 1966.

Middleman, S. *The Flow of High Polymers*. John Wiley & Sons, Inc., New York, 1968.

Nielsen, L. E. *Mechanical Properties of Polymers*. Reinhold Publishing Corp., New York, 1962.

Odian, G. *Principles of Polymerization*. McGraw-Hill Book Co., New York, 1970.

Rodriguez, F. *Principles of Polymer Systems*. McGraw-Hill Book Co., New York, 1970.

Schildknecht, C. E., ed. *Polymer Processes*. John Wiley & Sons, Inc., New York, 1956.

Schmidt, A. X. and C. A. Marlies. *Principles of High-Polymer Theory and Practice*. McGraw-Hill Book Co., New York, 1948.

Skelland, A. H. P. *Non-Newtonian Flow and Heat Transfer*. John Wiley & Sons, Inc., New York, 1967.

Sorenson, W. R. and T. W. Campbell. *Preparative Methods of Polymer Chemistry*, second ed. John Wiley & Sons, Inc., New York, 1968.

Tobolsky, A. V. *Properties and Structure of Polymers*. John Wiley & Sons, Inc., New York, 1960.

Wohl, M. H. Designing for non-Newtonian fluids. *Chemical Engineering*, Jan. 15, p. 148; Feb. 12, p. 130; March 25, p. 99; April 8, p. 143; May 6, p. 183; June 3, p. 95; July 1, p. 31; July 15, p. 127; Aug. 26, p. 113, all (1968). Also available as reprint no. 48.

INDEX